Industry 4.0 Topical Conference

Topical Conference at the 2023 AIChE Spring Meeting and 19th Global Congress on Process Safety

Houston, Texas, USA
12-16 March 2023

ISBN: 978-1-7138-7935-0

Printed from e-media with permission by:

Curran Associates, Inc.
57 Morehouse Lane
Red Hook, NY 12571

Some format issues inherent in the e-media version may also appear in this print version.

Printed with permission by Curran Associates, Inc. (2023)

For permission requests, please contact AIChE
at the address below.

AIChE
120 Wall Street, FL 23
New York, NY 10005-4020

Phone: (800) 242-4363
Fax: (203) 775-5177

www.aiche.org

Additional copies of this publication are available from:

Curran Associates, Inc.
57 Morehouse Lane
Red Hook, NY 12571 USA
Phone: 845-758-0400
Fax: 845-758-2633
Email: curran@proceedings.com
Web: www.proceedings.com

TABLE OF CONTENTS

DATA ANALYTICS AND SMART MANUFACTURING II

DATA ANALYTICS AND STATISTICS

DATA-DRIVEN AND HYBRID APPROACHES TO DEVELOPMENT OF NEW PRODUCTS I

DIGITAL TRANSFORMATION IN INDUSTRY 4.0 I

DIGITAL TWINS, AUGMENTED REALITY, AND ADVANCED SIMULATION

EMERGING TECHNOLOGIES IN DATA ANALYTICS

INDUSTRY 4.0 / FUELS AND PETROCHEMICALS DIVISION JOINT PLENARY

INVITED TUTORIAL SESSION - APPROACHES IN DATA ANALYTICS I

POSTER SESSION: INDUSTRY 4.0/ANALYTICS AND AI

ROBOTICS, CYBERPHYSICAL, AND AUTONOMOUS SYSTEMS

SENSORS AND ANALYZERS FOR ONLINE MEASUREMENTS

INVITED TUTORIAL SESSION - APPROACHES IN DATA ANALYTICS II

VENDORS' PERSPECTIVE (SESSION SPEAKERS INVITED)

OPTIMIZATION AND MACHINE LEARNING IN CHEMICAL MANUFACTURING

DIGITAL TRANSFORMATION IN INDUSTRY 4.0 II

CYBERSECURITY IN THE CHEMICAL INDUSTRY II

DATA ANALYTICS AND SMART MANUFACTURING I

DATA-DRIVEN AND HYBRID APPROACHES TO DEVELOPMENT OF NEW PRODUCTS II

REAL-TIME APPLICATIONS OF DATA ANALYTICS AND MACHINE LEARNING I

149a Exploiting Knowledge When Learning Reaction Rates with Neural ODEs from Experimental Observations

Farshud Sorourifar, You Peng, Ivan Castillo, Linh Bui, Juan M. Venegas

In this work, we investigate the effects of incorporating structural knowledge on the accuracy of a trained NODE (Neural Ordinary Differential Equations). Using a simulation case study, we investigated this effect when perfect knowledge is available, and found that such information does appear to improve the final model. We further investigated this phenomena on Grignard data set, and found incorporating structural assumptions based off the expert models reduced the final model's performance. Thus, incorporating structural information appears to reduce the expressiveness of a NODE. When the information accurately reflects the underlying mechanics, the reduction in expressibility converges toward the truth. However, inaccurate information may reduce expressibility such that an accurate model cannot be expressed. Future work may investigate combining an interpretable model formulation, such as SISSO into the NODE-style framework. This can be accomplished by training an interpretable model via the adjoint sensitive method used in the NODE framework.

149b Rapid Capability Analysis Using Discrete-Rate Simulation

Scott J. Bury, Andrew Moeller

This talk will highlight the use of Discrete-Rate simulation (DRS) technology for process analysis with the objective of determining the capability of the process to meet variable demand patterns. DRS is a simulation technology that bridges the gap between discrete time (e.g. Euler integrated) and discrete event (e.g., truck loading and unloading) simulation[1]. This enables efficient simulation of hybrid systems that have both continuous flow and discrete events such as a continuous chemical process feeding multiple storage tanks and from those tanks, distributing the product to pails, drums, ISO containers and tankers. We will describe two examples the first one is primarily a quality assurance problem and the second is understanding production capability to meet the expect change in product demand.

[1] Cecile Damiron, D. Krahl. "A GLOBAL APPROACH FOR DISCRETE RATE SIMULATION" (2014) Proceedings of the 2014 Winter Simulation Conference A. Tolk, S. D. Diallo, I. O. Ryzhov, L. Yilmaz, S. Buckley, and J. A. Miller, eds.

149c External Risks Created By Reliance on Open-Source Artificial Intelligence Libraries

Joshua Holtz

The most powerful open-source AI toolkits such as Pytorch and Tensorflow are developed by Meta (formerly Facebook) and Alphabet (formerly Google) respectively. The open-source AI community, including the chemical industry, have largely come to rely on these toolkits to develop high impact AI pipelines today. It is clear why these tech companies would want to develop these tools for in house use but not so clear as to how long they plan to make the toolkits available to the public or why they do in the first place.

The questions I want to explore are these:

1. What risks are we facing by relying so heavily on these toolkits?
2. Should we mitigate those risks?
3. How can we mitigate those risks?

17b Towards a Discipline of Industrial Process Analytics: From Case-Based Problem Solving to a Systematic Body of Knowledge

Marco Reis

The use of data for supporting inductive reasoning, operational management, and process improvement, has been a driver for progress in modern industry. Many success stories have been shared on the successful application of data-driven methods to address different open challenges, across different industrial sectors. The current and future importance of Industrial Process Analytics is therefore undeniable and is becoming widely considered as a competency that should be internalized and integrated into the company's knowledge lake and skills set. But isolated success stories, even when abundant, are of limited value if the only way to make use of them is by replication on closely related scenarios. It is the mechanisms that have conducted to such successes, not the final solutions themselves, that really matters when it comes to creating sustainable value from Industrial Process Analytics. The recent advances on AI/ML technology in the fields of image& video analysis and natural language have generated many new methods and tools, spiking the interest of the research community to explore their application outside these domains, namely in the chemical, food, biotechnological, semiconductor, and pharmaceutical industries, among others. But this boost in activity has also increased the difficulty of understanding the multiple underlying rationales for applying them, other than the mere curiosity of "to see what comes out" (still valid, but arguably inefficient). Furthermore, it is often difficult to assess the added value of using these new methods, as many times they are not rigorously compared with conventional solutions presenting state-of-the-art performances. Therefore, it is now (the) time, I believe, to work on the formal structuring of a discipline of Industrial Process Analytics, and establish a solid methodological ground for research and teaching in this field. This is essential for undergraduate engineering students to learn basic inductive skills and to carry them to industry, where they become naturally part of the company's knowledge lake. When we achieve that, then the transformative power of Industrial Process Analytics will be more than a set of isolated successes, but the result of implementing the principles and methodologies from a consolidated and structured discipline.

17a Leveraging Data Driven Connected Worker Technologies to Empower Your Workforce and Drive Operational Excellence

Constantyn Chalitsios

Westlake Vision of Digital Transformation in Manufacturing includes connecting data across systems and operations functions enabling data driven insights to enhance operational excellence that result in improvements to revenue, cost reduction, capital position, and sustainability. Integral to our Digital Transformation journey are digital technologies such as electronic rounds and augmented reality to connect field workers across our operations. Successful deployment depends on the business case showing economic value add, and overcoming the cultural barriers to adopting connected worker technology. We have taken a people-first approach to digital transformation and engage site participation in developing proof of concepts for our field workers. In collaboration with the sites we gathered their needs for a system architecture required to implement a data driven connected worker operating environment, and an implementation roadmap.

28a A Hybrid Modeling Approach for Catalyst Monitoring and Lifetime Prediction

Linh Bui, Ivan Castillo, Birgit Braun, You Peng, Mark Joswiak, Ailene G. Phillips, Jin Yang, Justin Rose, Daniel Hickman

Summary

In this work, we present a hybrid modeling approach combining a fundamental reactor model with a data-driven empirical model to predict the lifetime of a commercial heterogeneously catalyzed reactor system. The fundamental model connects the inlet conditions, temperature profile, and conversion of the reactor with the catalyst activity – the main KPI of the process. The empirical model, in turn, relates the catalyst activity with the process variables and catalyst age, allowing for prediction of future performance. The state of the model is updated automatically with daily new plant data using filter algorithms. The model provides a tool for catalyst health monitoring, process disruption monitoring, catalyst lifetime prediction, and turn-around planning.

Introduction

Catalyst deactivation is a major cause for loss of production and unplanned turnarounds in chemical plants. The capability to predict the performance metrics of the reactor, such as production rate, conversion, product yield, etc., at different conditions in the future using catalyst activity as a proxy is highly desired. However, the inability to measure the catalyst activity on a commercial plant scale, the numerous and complex fundamental mechanisms for catalyst deactivation, and the years-long time scale of relevant deactivation modes (compared to days-long typical lab experiments) render it difficult to construct a first principle model for catalyst deactivation that is relevant for commercial chemical plants.

Hybrid models combine engineering-based fundamental model with data-driven empirical model to improve the model's ability to explain real data while preserve its ability to make meaningful and explainable predictions for hypothetical, unobserved inputs. For an in-series hybrid model, the empirical portion calculates internal parameters for the fundamental model which, in turn, solves engineering equations to compute the outputs of the system. Kinetic parameters [1] or reaction rates [2] have been calculated from neural networks before being used in mass balance equations to solve for the conversion and yields of chemical reacting systems. In this work, we demonstrated the methodologies to construct, validate, and deploy an in-series hybrid model to estimate historical catalyst activity and predict catalyst lifetime and future performance of commercial reactors [3].

Methods

The first half of the constructed in-series hybrid model was a reactor model that contained the fundamental mass and energy balance equations, thermodynamics, and kinetics of the reacting system and used the historical inlet conditions and indicators of the catalyst activity, such as conversion, yield, temperature profile, etc., to calculate the historical catalyst activity. The partial-least-squares model then found an empirical correlation between the historical catalyst activity and the process variables in the chemical plant and allowed prediction of the catalyst activity in hypothetical reaction conditions. In tandem, the two fundamental-empirical models enabled the projection of the reactor performance in the future. The model was calibrated with process data from a commercial plant with 24 catalyst charges and 18 years of continuous operation, resulting in 370 variables and over 18,000 data points.

Results and Discussions

The calculated historical catalyst activity showed the expected decreasing trend with operation time and cumulative reactant feed. The empirical model successfully captured the catalyst activity as a function of seven process variables and achieved a goodness-of-fit of 87% and goodness-of-prediction of 61%, demonstrating its ability to predict the catalyst activity for unseen data. A 100-day look-ahead validation scheme was employed in which the empirical model predicts the catalyst activity 100 days from the date-of-interest and the fundamental model predicts the reactor conversion. Quantitative agreement was observed between the historical data and the prediction from this validation scheme, further highlights the ability of the hybrid model to predict key performance metrics of the commercial reactor. A projection tool from this model allowed the user to monitor the historical catalyst activity, predict the future activity, and estimate the timing of the next required turn-around and the remaining production.

Conclusion

This work demonstrated a framework for constructing, validating, and deploying hybrid fundamental-empirical models to monitor and predict the catalyst activity of commercial reactors, from there improved the reactor's performance, optimized the turn-around planning of the chemical plant, and increased overall production and efficiency.

References

1. Azarpoura et al. *Chemical Engineering Research and Design*, vol. 117, pp. 149-167, 2017.
2. Saraceno, S. Curcio, V. Calabrò and G. Iorio, *Computers and Chemical Engineering,* vol. 34, pp. 1590-1596, 2010.
3. Bui et al. *ACS Eng. Au*, vol. 1, pp. 17-26, 2022

28b Physics-Informed Deep Learning for Prediction of Thermophysical Properties: Normal Boiling Point

Mohammad R. Babaei, Frank Mtetwa, Ryan Stone, Thomas Knotts IV, John Hedengren

Chemical thermophysical properties are needed for chemical handling, design of production and storage facilities, separations, and manufacture. The Design Institute for Physical Properties (DIPPR) was created in 1978 (under the direction of the American Institute of Chemical Engineers) and is the best source of critically evaluated thermophysical, safety, and environmental properties. When experimental data for a chemical is not available, DIPPR predicts the values for properties of that chemical. Creating more accurate and broadly applicable prediction methods for thermophysical properties is an area of constant research. One example of a relevant thermophysical property is normal boiling point (NBP).

A set of 1600 compounds is utilized for training, validating, and testing a Physics-Informed Neural Network (PINN) to improve Normal Boiling Point (NBP) based on the group contribution methods [1]. Physics-informed deep learning seeks to improve predictive accuracy by incorporating physics-based information with machine learning. Standard artificial neural networks have a known weakness to extrapolation potential when used outside the training region. An artificial neural network not only produces a prediction, but also a self-assessment of uncertainty. The contributions to the parachor should be strictly additive in nature as it represents a volume of space a molecule occupies. However, the approaches mentioned above have several groups which are negative, suggesting a suboptimal optimization relating parachor to the groups. We recently combined machine learning (ML) with a physics-based constraint to achieve better predictions than any previous method for surface tension [2]. This presentation outlines additional progress with NBP with a comparison to other leading prediction methods. For compounds with Tb > 600 K, the PINN model yields 24.3◦C Mean Absolute Error (3.6% Mean Percentage Error), while this value for the Joback method is about 79.0◦C MAE (or 12% MPE). Across the results, predicted NPB for compounds containing silane and imine families are less accurate. NBP results further demonstrate that physics-based constraints with machine learning produce significant improvements in prediction methods for the thermophysical properties that are crucial in the field of chemical engineering.

References Cited

[1] Ericksen, Wilding, W.V., Oscarson, J.L., and Rowley, R.L., Use of the DIPPR Database for Development of QSPR Correlations: Normal Boiling Point, J. Chem. Eng. Data 2002, 47, 5, 1293–1302, July, 2002 DOI: 10.1021/je0255372

[2] Knotts, T., Hedengren, J.D., Babaei, M.R., Physics-Informed Deep Learning for Prediction of Thermophysical Properties: The Parachor Method for Surface Tension, AIChE Annual Meeting, Phoenix, AZ, Nov 13-18, 2022.

28c A Hybrid Approach for Fouling Monitoring and Forecasting with Application to an Industrial Heat Exchanger

Joel Sansana, Ricardo Rendall, Ivan Castillo, Luc de Bruijne, Jonathan Huggins, Ailene G. Phillips, Marco Reis

Heat exchanger fouling in ethylene oxide (EO) plants is a safety hazard that can lead to unplanned outages and temperature excursions or decomposition events, therefore constituting a major risk for safe operations (Hess & Tilton, 1950; Cavitt, 1983). Fouling of the heat exchangers reduces energy recovery and increases pressure drop. The increase in pressure drop eventually leads to added expenditures in utilities to maintain the process flows at a target design level. Furthermore, fouling deposits originate localized heat transfer dead zones where hotspots can form leading to strong temperature excursions and decomposition events (Crocco et al., 1959). All these effects are detrimental to the plant operation and should be managed as closely as possible to secure the stability, economy, and safety of the process. However, in most circumstances, fouling is not directly observable or measured, and the only alternative available is to infer it from the analysis of collected data or field measurements, usually in a rather ad hoc fashion, depending on the operators' experience (Sundar et al., 2020). Therefore, the development of a framework for advanced monitoring and forecasting of heat exchanger fouling is both opportune and important to improve the reliability safety of the operation (Trafczynski et al., 2021).

We propose a hybrid approach, where knowledge-based feature generation is integrated with data-driven methods, to forecast a key performance indicator (KPI) that acts as a fouling surrogate (Diaz-Bejarano et al., 2020). Knowledge-based feature generation allowed to monitor the evolution of fouling over time and enabled the use of off-the-shelf data-driven forecasting methods. Among the KPIs tested, two were selected to act as fouling surrogates. We advise monitoring them in tandem with the Reynolds number and the ratio $\Delta T/\Delta T_{ml}$ as KPIs because the first one reflects the flow conditions in the heat exchanger while the ratio provides information on the heat transfer phenomena. The developed forecasting model, a time series multiple regression model, was capable to predict the KPI one-month ahead with a testing accuracy of R^2=0.7. Furthermore, we showed that long-term forecasting is also possible with this model, always with the caveat that the furthest we look into the future, the more uncertainty the forecast carries. Nevertheless, the model can still be applied for process optimization and maintenance scheduling.

In the future, we believe it will be opportune to further develop the model as well as the optimization framework by including uncertainty quantification on the inputs and in the model. In this way, more robust decisions can be made that consider process safety, reliability, and profitability.

References

Cavitt SB. Ethylene oxide production. US Patent 4,374,260 (Feb. 15, 1983).

Crocco L, Glassman I, Smith I. Kinetics and mechanism of ethylene oxide decomposition at high temperatures. *The Journal of Chemical Physics*. 1959;31.

Diaz-Bejarano E, Coletti F, Macchietto S. A model-based method for visualization, monitoring, and diagnosis of fouling in heat exchangers. *Industrial & Engineering Chemisty Research*. 2020;59.

Hess L, Tilton V. Ethylene oxide-hazards and methods of handling. *Industrial & Engineering*

Chemistry. 1950;42.

Montgomery DC, Peck EA, Vining GG. Introduction to Linear Regression Analysis. 5th Edition, John Wiley & Sons, Inc., Hoboken, New Jersey, 2012.

Sundar S, Rajagopal MC, Zhao H, Kuntumalla G, Meng Y, Chang HC, Shao C, Ferreira P, Miljkovic N, Sinha S, Salapaka S. Fouling modeling and prediction approach for heat exchangers using deep learning. *International Journal of Heat and Mass Transfer*. 2020;159.

Trafczynski M, Markowski M, Urbaniec K, Trzcinski P, Alabrudzinski S, Suchecki W. Estimation of thermal effects of fouling growth for application in the scheduling of heat exchangers cleaning. *Applied Thermal Engineering*. 2021;182.

85a Cyber Risk Assessment Methodologies - Compare and Contrast Different Approaches ISA 62443, INL Cce

Marco Ayala

Today's automation systems are much more highly connected than they were at the start of this millennium. Advances is technology, the cost of computing and the speed of communication in just the last seven years have brought industries higher performance, optimization, visibility, advanced system diagnostics with high reliability and availability fault tolerance, analytics and production yields that unquestionably have revolutionized all industry sectors. Convergence of site automation system technologies and traditional enterprise IT systems have become more prevalent and provide opportunities such as the remote mobile worker, integration into MES and ERP systems and remote access for troubleshooting and support. While this has benefits it also brings inherent risk to plant operating environments.

This talk will review and compare the standards-based (ISA 62443-3-2) approach and the Consequence-driven Cyber-informed Engineering (Idaho National Labs CCE) methodologies.

85c *Cyber Security Visibility in OT: The Active Approach*

Randy Conner, Rutger Hendricks

Engineers and cyber security personnel face a difficult challenge getting a good understanding of the risks that exist in their OT environment. Traditional methods for performing security assessments, such as penetration testing, site questionnaires or passive network monitoring all have their significant drawbacks.

Join Nouryon in their journey to explore what it takes to build a better risk picture of their industrial control systems, and discover if the benefits of active monitoring outweigh the risks.

70a Development of Advanced Image Analysis Algorithms for Quantitative Evaluation of Exterior Coating Degradation

Xingyu Zhou, Michael Linsen, Stephanie Long, Sun Hye Kim, Kevin Henderson, Jonathan DeRocher, Yicheng Hu

Coating durability, an ability to maintain performance for the intended service period, is an important performance metric for exterior coatings. Traditionally, performance attributes related to durability have been rated by a researcher using a set of guidelines, a method prone to subjectivity. To overcome this challenge, the Dow Coating Materials team has developed a novel workflow to allow quantitative analysis of coating degradation. With the use of new digital image capture technology under a controlled setting and advanced image analysis algorithms, this work provides a robust and quantifiable assessment of coating performance over time. This talk will focus on several uses cases to illustrate the holistic workflow such as: reliably capturing images of exterior coatings, data pre-processing, development of image analysis algorithm, and algorithm deployment. We will also demonstrate the use of both traditional statistical methods and deep learning methods and illustrate how these techniques can be leveraged in different scenarios.

70c Solubility Prediction of Industrial Chemicals: Feeding Graph Neural Networks with Physics-Based Simulations Data

Ravindra Aglave, Shikha Mallick, Francesca Boioli, Panos Petris, Peter Mas

The solubility of industrial chemicals dictates performance, stability, and process development. To assess this important material property in a fast and cost-effective way, digital solutions can be utilized. In the literature there are plenty of physics-based simulation and data-driven approaches. In our computational framework, we combine the best of both worlds to provide a complete solution to this industrial need.

As a first step, a limited set of molecules from a specific chemical family (or families) is selected, and their solubility is calculated using molecular simulations. More specifically we estimate the Hansen solubility parameters[1] using a fast Molecular Dynamics technique, where the simulation box is successively compressed and de-compressed to reach faster in correct density and thermodynamic equilibrium.[2] The solubility parameters are defined by the intermolecular energy of the simulation box.

We subsequently use the simulation data to train a Graph Neural Network (GNN). The adopted machine learning model is very meticulously selected. Traditional machine learning algorithms and deep learning algorithms[3] were also tested and provided high accuracy results, but molecular descriptors had to be generated to train the model against them (i.e. asphericity, molecular weight, number of hydrogen bonding donors/acceptors, radius of gyration, etc). The advantage of GNNs is that the complex chemical structure of a molecule can be naturally represented by a graph, defined as the ensemble of the connectivity relationships between a set of nodes (atoms) and a set of edges (bonds).[4] That allows us to immediately train against the target solubility parameters without additional calculations.

Our hybrid protocol provides us with a very fast and accurate method for solubility calculation of industrial chemicals. The fast Molecular Dynamics technique should be performed only once for a specific chemical family. Once the GNN is trained then it automatically predicts the representative value for any other chemical of the same family. Such framework is more accurate than any Quantitative-Structure-Property-Relationship[5] or data approach, while much faster than any particle simulation technique.

(1) Hansen, C. M., J. Paint Technol. 1967, 39, 511

(2) Belmares, M.; Blanco, M.; Goddard, W. A.; Ross, R.; Caldwell, G.; Chou, S.-H.; Pham, J.; Olofson, P. M.; Thomas, C., J. Comput. Chem. 2004, 25: 1814–1826

(3) Deng, T.; Jia, G.-Z., Molecular Physics, 2020, 118:2

(4) Lee, S.; Lee, M.; Gyak, K.-W.; Kim, S.; Kim, M.-J.; Min, K., ACS Omega 2022, 7, 14, 12268–12277

(5) Chinta, S.; Rengaswamy, R., Ind. Eng. Res. 2019, 58, 8, 3082–3092

70b Deep Learning Applications in Advanced Analytic Approaches of Big Data

Manuel Rodriguez, Ismael Díaz Moreno, Hamed Aardakani

Recent versions of industries known as Industry 4.0 and Industry 5.0 are set up on wide ranges of the cutting-edge technologies such as Internet of Things (IoT). IoT is considered as a dynamic global network infrastructure with self-configuring capabilities based on standard and interoperable communication protocols where physical and virtual 'Things' have identities, physical attributes, and virtual personalities and use intelligent interfaces [1] [2]. IoT is among those technologies that facilitates data flow lead to making massive volume of data, known as Big Data. Several definitions and explanations are suggested to address Big Data, but it mainly refers to massive datasets that usual hardware and software cannot collect, process, and analyze it [3] [4].

Performing advanced analytic tasks would become more complicated and demanding when they have to cope with inherent Big Data characteristics. Analytic approaches that deal with Big Data aim to identify and predict current and future patterns and trends to describe process situations. Most conventional machine learning algorithms cannot detect complicated hidden patterns and trends in Big Data. In contrast, Deep learning (DL) algorithms have presented superb performance in satisfying this requirement through automating the learning procedure and providing complex, robust models. DL algorithms facilitate being confronted with intrinsically troublemaking aspects of Big Data, which derives from DL aptitudes for analyzing large volumes of data available with a high rate from multiple sources. DL algorithms can learn from unsupervised data, in addition to supervised data, which is a key feature for any algorithm to be applied in advanced analytics approaches.

DL was presented in 2006, however, the roots of the concept date back to the 1940s [5]. DL models are deeper variants of artificial neural networks (ANNs) with multiple layers, whether linear or non-linear. Each layer is connected to its lower and upper layers through different weights. Through attempting to imitate the hierarchical learning structure of human brain, DL algorithms automatically extract patterns, representations, and trends from data. DL, in fact, is a deep architecture of layers that apply nonlinear transformations to input to provide representations of it in output [6]. Two appropriate DL architectures for advanced analytical approaches of Big Data are Deep Belief Network (DBN) and Convolutional Neural Networks (CNN). DBN constructs a model using unsupervised and supervised techniques, consisting of an input layer, hidden layers, and an output layer, while CNN is a supervised technique that has multiple hierarchical layers [7] [8].

Enhanced DL architectures assist advanced analytic approaches to discover meaningful Big Data patterns for making decisions, predictions, and conclusions. As an instance, it is proven that the parameters of large-scale DL models, especially with fully connected layers, are of high redundancy. Therefore, some methods, such as Hashing Trick, have been presented to improve the training efficiency by significantly compressing the parameters without a noteworthy accuracy drop. In addition, some software such as DistBelief is developed to train large-scale DL models in multiple machines in parallel. Through partitioning a large deep model, it significantly improves the training efficiency [5].

During the last decade, because of DL algorithm success in various applications and according to Google Trends, DL have received more attention compared to other popular machine learning algorithms. In this regard, this research investigates recent advancements in DL algorithms and their applications in advanced analytic approaches for Big Data analyzing.

Keywords: Deep learning, Analytics, Big Data, Industry 4.0, Industry 5.0

References

[1] L. Da Xu, E.L. Xu, L. Li, Industry 4.0: State of the art and future trends, International Journal of Production Research. 56 (2018) 2941–2962. doi:10.1080/00207543.2018.1444806.

[2] B. Chen, J. Wan, L. Shu, P. Li, M. Mukherjee, B. Yin, Smart Factory of Industry 4.0: Key Technologies, Application Case, and Challenges, IEEE Access. 6 (2017) 6505–6519. doi:10.1109/ACCESS.2017.2783682.

[3] H. Ardakani, H. Hemmati, M.R. Hernandez, Data Driven Process Monitoring, from Fault Detection and Diagnosis Points of View, in Industry 4.0 Context, 32 European Symposium on Computer Aided Process Engineering, Elsevier, 2022: pp. 1033–1038. doi:https://doi.org/10.1016/B978-0-323-95879-0.50173-9.

[4] M. Chen, S. Mao, Y. Zhang, V.C.M. Leung, Big Data challenges and technologies, related future prospects, 2010. http://www.springer.com/series/10028.

[5] Q. Zhang, L.T. Yang, Z. Chen, P. Li, A survey on deep learning for big data, Information Fusion. 42 (2018) 146–157. doi:10.1016/j.inffus.2017.10.006.

[6] M.M. Najafabadi, F. Villanustre, T.M. Khoshgoftaar, N. Seliya, R. Wald, E. Muharemagic, Deep learning applications and challenges in big data analytics, Journal of Big Data. 2 (2015) 1–21. doi:10.1186/s40537-014-0007-7.

[7] X. Wang, Y. Zhao, F. Pourpanah, Recent advances in deep learning, International Journal of Machine Learning and Cybernetics. 11 (2020) 747–750. doi:10.1007/s13042-020-01096-5.

[8] B. Jan, H. Farman, M. Khan, M. Imran, I.U. Islam, A. Ahmad, et al., Deep learning in big data Analytics: A comparative study, Computers and Electrical Engineering. 75 (2019) 275–287. doi:10.1016/j.compeleceng.2017.12.009.

103a Total Dissolved Solids Forecast and Risk Assessment in Midland Wastewater Treatment Plant

Caterina Rizzo, You Peng, Shane Bennett

Discharge of total dissolved solids (TDS) represents a limiting factor for the wastewater treatment plant (WWTP) at Michigan Operations in Midland as permits regulate it. These compliance requirements involve mass loading and concentration balance downstream in the river. Whereas the first is easier to control and predict, the latter heavily depends on uncontrolled conditions, such as weather and background river TDS. Additionally, the TDS concentration significantly affects the compliance requirements for Whole Effluent Toxicity (WET). Intuitively, this aspect substantially affects the normal operation of those production processes that discharge wastewater with high TDS concentration.

The control strategy consists of utilizing a pond for equalizing and optimizing the discharge flow as the river and weather conditions allow. Furthermore, coordination with sender plants is critical to ensure appropriate management of incoming streams to the WWTP. and involves coordination with the sender plants. In this context, the uncertainty of the uncontrollable variables poses a severe limitation and quantifying the risk of a compliance violation can improve decision-making on the control strategy and production coordination. The risk of exceeding discharge permits can be quantified using stochastic simulations. Stochastic simulations allow recreating all the possible outcomes that drive the control strategy decision by propagating the uncertainty of the input variables via deterministic models to obtain a probability distribution of the output of interest.

The analysis of the extensive historical records of production, weather, and river conditions can help identify the variables that exert a strong impact on the TDS discharge, understand daily fluctuations, and evaluate the influence of environmental conditions. Moreover, data analysis allows randomly generating inputs from well-defined probability distributions over the domain of interest, considering seasonality effects, correlations, and time-series dependencies. This study is also helpful for comparing and validating the deterministic models shaping the relationship between inputs and outputs. This setup allows modeling interdependent and non-linear relationships governed by mass balances.

The models were implemented in an interactive and friendly user-interface web application built in RShiny, allowing WWTP personnel to examine multiple scenarios and quantify their associated risks and it provides a complete solution for long-term and short-term risk assessment.

103b Application of Functional Data Analysis (FDA) for Process Optimization

Swee-Teng Chin, Joel Sansana, Ricardo Rendall, Matthew Malloure, Ivan Castillo, Zhenyu Wang, Jinyang Chen, João P. L. Coutinho, Brian Clark

Functional Data Analysis (FDA) is an emerging field in statistics that allows for characterizing data over a continuum. Recently, FDA has gained a lot of attention in the literature and it has been successfully applied to data across multiple businesses in Dow R&D, but it hasn't been applied to manufacturing data to this point. This motivates the interest of applying FDA in manufacturing applications, where analyzing time-series data is increasingly more common, especially in modeling and optimization of batch processes. Previous work had revealed some technical challenges that needed to be addressed to apply FDA to analyze manufacturing data. In this presentation, we will demonstrate how to deal with such gaps and apply FDA to manufacturing data for process optimization, deliver a comparison of FDA with existing methods, and recommend optimum conditions for applying FDA in modeling and optimization.

103c Analysis of Foaming in Batch Fermentation Processes Using Observation-Wise and Batch-Wise Unfolding Partial Least Square (PLS) Approaches

Xuan Dung James Nguyen, Niket Sharma, Y. Liu, Christopher McDowell

This work presents an essential application of multivariate data analysis to fermentation process monitoring and quality enhancement, especially the prediction of the occurrence of foaming in fermenter operation. Foaming is a phenomenon that frequently occurs inside fermenters and reduces their effective working volume, resulting in a decrease in product yields. This study performs a constructive quantitative analysis of predicting foaming occurrence using a technique commonly used in chemometrics called partial least square (PLS), extended to Multiway PLS (MPLS) for three-dimensional data matrices. At the same time, we examine the relationships and correlations between the product quality measurements and the independent process variables. Consequently, we try to analyze those relations to the foaming problems in the fermenter. In addition, we also compare and contrast two unfolding methods, Observation-Wise Unfolding (OWU) and Batch-Wise Unfolding (BWU). To our knowledge, no published study has demonstrated the applications of multivariate data analysis in the foaming prediction of the fermentation processes and the comparison between the two aforementioned unfolding techniques. Concerning the OWU method, we use two approaches of MPLS, Kernel Partial Least Square (KPLS) and Dynamic Partial Least Square (DPLS), where we consider the exhaust differential pressure as the dependent variable. With respect to the BWU method, we refer to the product quality variables as dependent variables and apply PLS to correlate the dependent variables to process variables. The OWU KPLS/DPLS approaches show that the exhausted differential pressure positively correlates with volume and pH while negatively correlates with dissolved oxygen. In the BWU approach using PLS, we conclude that foaming has a negative impact on the quality variables. We recommend the BWU approach when analyzing the differences among batches and monitoring the product quality, and we prefer the OWU approach when the dependent variable is time-dependent. Thus, in this study, we recommend the OWU approach for the comprehensive analysis of foaming.

116a Making the Most of Small Data

Michael Heiber, Christopher Farrow

For many traditional innovation-driven organizations, scientific data is generated to answer specific immediate research questions and then archived to protect IP with little attention paid to the future value of reusing the data to answer other similar or tangential questions. Data is essentially a side product of R&D and not viewed as a primary output. As a result, important experimental process details and implied contextual information are often not recorded. In addition, data is often not formatted in a consistent and well-structured manner, which makes it difficult and expensive to parse large volumes of historical data files that may be archived in a network drive or data lake. Finally, the experimental workflows that produce this data are manual and often require coordination between multiple teams. Manual sample preparation and handoff between labs, manual data transfer between computers, manual raw data analysis on instrument computers, and waiting for external analytical measurement requests to be completed all make new data generation very slow and expensive. Altogether, this leads to R&D labs having surprisingly small datasets that are actually clean enough and complete enough to serve as training data for a machine learning model.

Faced with their ‘small data’ situation, researchers and managers often feel that they may not yet benefit from pursuing data-driven approaches to new product development. They are not sure what can be done given the current state of their data, and they are not sure how to efficiently gather more data to alleviate the issue. They may even think it is impossible to pursue this approach given the current cost of new data generation. It is not uncommon to feel stuck without a clear path forward. Other organizations push forward a high-level vision and implement data platforms and hire data science and engineering teams that then soon struggle to generate value due to the unique challenges inherent to scientific small data problems.

At Enthought, we have tackled many small data challenges in materials and chemicals product development and have employed multiple strategies for getting the most value out of small data to meet strategic innovation goals. There is not a one-size-fits-all solution, but in this talk, we’ll present practical tips for how teams can make the most of what they have and set a course towards continuous improvement. We’ll discuss how teams can get started with little to no data at all and how they can leverage existing domain knowledge to get further with less data through well-crafted experimental designs, feature engineering, informed model constraints and priors, and improved data quality. We’ll also discuss how to assess existing data generation workflows and prioritize workflow improvements that will accelerate new data generation and improve data quality using software tools to streamline data labeling tasks and to automate or assist users with raw data analysis. We’ll then conclude with a discussion about workflow innovation where domain experts explore screening approaches that leverage theory and simulation together with proxy measurements to accelerate learning at lower cost.

116b Leveraging Small Data for Better Chemistry

Daniela Blanco

Even though over 95% of goods around us come from the business of chemistry, developing new materials, formulations, and chemical production processes is a time-intensive and expensive challenge. Complex optimization efforts are commonly required to enhance production performance while avoiding harsh reaction conditions, high costs, polluting effects, and other limitations.

Data-driven machine-learning (ML) offers the possibility to dramatically reduce the resources needed for reaction optimization through supervised-learning models that predict reaction performance, catalytic behavior, and optimal reaction conditions. However, most ML algorithms require large datasets (i.e. 100's-1000's of points) to provide accurate predictions, while experimental data collection is expensive in human and material resources. The challenge of building large datasets can be averted with Bayesian Optimization frameworks that guide experimental campaigns through informed and smart decisions. Predictions are thus based on carefully selected training data, which ultimately reduces the number of experiments required to achieve reliable understanding of variable effects and reaction behavior.

Addressing fundamental questions regarding the limitations of a modified Bayesian Optimization approach, we elucidate a pathway for the design and implementation of intelligent experimental campaigns using an organic electrocatalytic cross-coupling transformation as a model reaction. The approach has been further validated on catalytic biomass conversion, dehydrogenation processes, and mechanocatalytic transformations, among others. Working at the intersection of ML and chemistry, we created SuntheticsML, an online, user-friendly, ML platform that leverages very small data (starting with only 5 data points) to enable the development of new chemical processes, materials, and formulations up to 15 times faster
without the need of coding or ML skills.

The insights from this work innovate on the future of reaction engineering, material design, and process optimization while lowering the barrier to the implementation of new chemical technologies. SuntheticsML accelerates lengthy and complex traditional optimization campaigns, harnessing the power of ML and small data to fast-track innovation, sustainability, and digitalization in the chemical industry, further reducing development waste, emissions, and resource consumption by up to 95%

116c Guided Experimental Design for Modeling Reaction Rate Expressions

Byanne Malluhi, Mamoun Al-Rawashdeh, Costas Kravaris, Hazem Nounou, Mohamed Nounou

Sequential experimental design uses knowledge from previous experiments to guide what experiments to run next. In the literature, this goes by many names; *optimal experimental design*, *adaptive experimental design*, and *goal-oriented experimental design*. Guided or goal-oriented experimental design methods help achieve the target cheaply and quickly. Recently, there has been a surge in using Bayesian optimization (BO) for guided experimental design. BO uses non-parametric regression to model an objective, and sequentially guides the optimization by maximizing an acquisition function. At every iteration, BO determines the next evaluation point and tries to sample in areas where the uncertainty is high (for explorative purposes) and in areas close to where the current best is (for exploitive purposes).

In this work, the experimental objective is obtaining accurate and reliable models for reaction rate expressions. BO convergence rates will be evaluated for different cases. The effect of increasing the dimension of the input variables, level of noise, and number of hyperparameters of the kernel functions on the BO convergence rates will be studied. Furthermore, BO becomes computationally and statistically inefficient for high dimensional systems. Developing BO methods that work well in higher dimensions is thus of great practical and theoretical interest. A large body of literature has been devoted to mitigate the challenges of high dimensional BO problems. Several of which exploit the intrinsic lower dimensionality of the objective function, and perform BO in the lower dimensional space, followed by an inversion back to the high dimensional space. In this work, the problem of higher dimensions will be addressed for the reaction modeling application where a limited number of initial samples creates a unique dimensionality reduction problem.

54b Capital Asset Lifecycle Management (CALM) for Chemical Industry.

Chandra Siddamreddy, John Nixon, Mallik Manem

Capital Asset Lifecycle Management (CALM) addresses the challenges of greenfield and brownfield capital projects, such as rampant delays and cost overruns, often caused by lack of system interoperability, low labor productivity and a failure to embrace digitalization. These challenges negatively impact the operational efficiencies of plants and facilities for decades. According to McKinsey & Company, 98% of megaprojects incur cost overruns or delays, averaging a cost increase of 80% and delaying delivery by 20 months. In addition lack of accurate information results in operational inefficiencies through the life of the plant which is anywhere from 30-50 years. These numbers are staggering and provide a ripe opportunity for the application of Siemens solutions. In this session, you'll learn about the industry's major trends and challenges while conveying the power of digitalization and demonstrating its value throughout the lifecycle of plant or facility. You'll gain an understanding of Teamcenter for Capital Asset Lifecycle Management and how, through the dissemination of engineering and asset data, it can be used to spur your customer's digital transformation.

54c Lessons Learned about Digital Transformation from Experienced Industry Professionals!

Jonas Norinder

In this panel discussion we will hear from three experienced industry individuals who will share their experience with and perspective on digital transformation and how it has shaped their companies as well as their own roles. In the discussion, we will put emphasis on the lessons they have learned as well as any suggested best practices they would recommend to other companies going through digital transformation journeys. Examples of topics to be discussed include how to manage people through change, how to prioritize projects to identify the ones that add the most value, examples of digital technology implementation and more.

Format: Panel discussion

Panelists:

- Entegris, speaker TBD
- Novozymes, Mads Torry Smith,
- Dupont, Duncan Coffey

Moderator:

- Jonas Norinder, Siemens

54a Digital Transformation of Process Design, Scale-up, and Control: Serving the Purpose

Nima Yazdanpanah, Thomas Eppinger, Jorge Lopez, Ravindra Aglave, Justin Hodges

The process digitalization has demonstrated its benefit at different levels and various applications. However, there is no one solution fits all. The digital transformation should be aligned with the application, resources, business objective, and maturity level of the process and enterprise. Depending on the applications, from R&D to post manufacturing, the tools, approaches, and implementations should be defined and developed according to the strategic planning, technoeconomic analysis, and value driven approach.

In this presentation, different tools and approaches and their interconnection will be discussed with case studies from equipment design and scale-up, manufacturing, and process control. The cases and platforms will demonstrate a holistic approach for the industry 4.0, that each tool will serve a specific need along the life cycle of the process/product.

132a Virtual Reality in the Context of Training Frontline Workers and the Process Industries

Iiro Esko

The purpose of this abstract is to share knowledge around the value of applying virtual reality when training process industries frontline workers.

The application of these technologies helps to attract talent and to maintain workforce through reuse of digital twins. Time to proficiency and quality improvements are directly enhanced and derived from leveraging gamification concepts in virtual training systems. The example case that will be presented is available for academia and industry professionals to visit in person at a US government funded innovation institute, where an example mobile worker and digital twin testbed has been built in collaboration with a leading US chemical company.

Moving from classrooms settings that center around slides, videos, pen and paper, and other traditional testing or training approaches to gamified virtual reality makes a difference both in learning velocity and sustaining best practices. Manual sequences are best learnt in an immersive but safe environment, learning by doing assembly work, such as mounting, dismounting, cleaning, draining, and the execution of maintenance tasks. Virtual training also serves different service providers both in the initial training as well as refresher trainings.

Ancillary savings around on-boarding, travel, information access, value chain collaboration, and training the trainers complete the operational benefits of increased safety, quality, productivity, and operations in general. Recommended for early career professionals, academia, industry professional both at plants and in the value chain.

132b Using Digital Utilities Twin Technology on a Refinery to Drive Carbon and Energy Reductions

Joana Brito, Hugo Carabineiro, Kunpeng Guo, Lorena Fernanda Dos Santos De Souza

Digital replicas of operating assets that combine plant data with high-fidelity process models are bringing a new level of decision support to operations of steam, condensate, fuel gas and power systems, otherwise known as 'utilities' systems. They provide many benefits to operators, from information on monitoring and 'soft-sensing' to advice on optimal set-points, diagnostics, and prognostics such as future maintenance interventions. At the heart of such shadowing systems is an always-current digital simulation model of the assets which updates itself periodically based on real-plant performance.

In the context of utility systems such as steam, condensate, electricity and power, digital utilities twin technology is now being implemented in many process industry applications. The associated benefits include: equipment monitoring to determine the real state of equipment; real-time 'soft-sensing' to provide up-to-date performance information which is either difficult or impossible to measure; forecasting to determine future performance such as maintenance intervals based on current equipment state and anticipated operation; operational optimisation to give advice to operators regarding set-points, diagnostics, prognostics and performance benchmarks; and finally 'what-if' analysis to anticipate how to operate for future/alternative operating scenarios.

One very specific focus for Utilities Twins is in the area of decarbonisation and energy reduction. The urgency in these areas has recently grown, with global warming and world events combined with the underlying drive by refining companies to find new ways to address their environmental commitments. Such twins have a significant contribution to make in this area.

The Digital Utilities Twins described in this paper use high-fidelity predictive mathematical models of the assets to exploit redundancy between model prediction and plant data and maintain themselves in an always-current state. This means that 'drift' in key parameters such as gas turbine efficiency, steam pressure levels, equipment fouling and so on is taken into account in all monitoring, forecasting and optimisation calculations, to provide reliable information for decision support.

This paper describes how Galp has been combining equation-based general-purpose modelling technologies with a next-generation digital application framework to form an environment for easy construction and application of fast, robust online solutions based on high-fidelity utility system models. The paper describes how the digital twin of the utilities system has been used to improve system performance across all areas of utilities in the refinery. The twin is linked to plant data systems, updating itself through machine-learning capabilities, validating actual performance and, where appropriate, identifying departures from normal operation. Beneficial operating changes are highlighted in web-based dashboards which operators can view and act upon. These dashboards have been developed through a collaboration between all stakeholders and they provide a comprehensive summary of utility system performance, all presented on a consistent basis. Visualisation of the twin is seen to be critically important to give operators greater insight and confidence to operate the process safely at the optimum point with a view to reducing green-house gas emissions and reducing energy consumption.

The Utilities Twin has also had a wider beneficial impact. It has been used to improve understanding and knowledge of energy consumption across the refinery within different stakeholders and been a focal point to make improvements in many different aspects of energy consumption. As such, the twin is delivering improvements, keeping the refinery utility system running optimally, and making a significant contribution to the refinery's decarbonisation and energy reduction goals.

132c Executable Digital Twin: Automation, Physics Based Simulation and Machine Learning

Ravindra Aglave, Iiro Esko

In the energy and process industries, physics-based digital twins are gaining prominence in virtual equipment development. Using simulation in the digital twin, engineers can determine optimal equipment designs before a unit goes into production.

Until recently, digital twins could not be deployed at the operational level, as they could be too complex or slow to react to the speed of operation. This is changing quickly. In this webinar, you'll learn how physics-based simulation lowers the barrier for executing the digital twins on the production floor.

Using simulation to understand complexity

Energy and chemical industry processes are complex. Using simulation to understand complexity creates opportunities to deploy the insights to your businesses' advantage. Physics-based simulation models, at varied length scales, allow engineers to capture this complexity and use it to explore design and operating spaces to find optimum conditions.

Deploying insights from simulation into operations

Deploy simulation insights into operations using executable digital twins. With digital twins, your engineers can determine the ideal design before the unit goes into production. Operations engineers can react to real-time product disturbances based on the deep insights gained during R&D.

Key requirements for physics-based simulation

The need to investigate problems across all lengths and time scales is one of the essential requirements for physics-based simulation. An engineering team may need a fit-for-purpose tool for a given scale. If multiple tools are required, integration must be seamless while offering a simple-to-use workflow.

87a Hybrid Modeling for Managing Process Changes in Chemical Processing Industries

Joel Sansana, Ricardo Rendall, Ivan Castillo, Caterina Rizzo, Birgit Braun, Leo Chiang, Marco Reis

The Chemical Processing Industries (CPI) has been challenged by shifting market demands, requiring flexibility to adapt to new scenarios readily. This transformation entails implementing new operational modes that are economically advantageous but still achieve the required quality and sustainability targets. Process changes are needed, for instance, when a new product grade is sought or when the optimal operating conditions shift significantly due to raw materials and utility costs. In this context, analytic technologies provide solutions to support process changes in CPI. Hybrid modeling has attracted a significant amount of research due to increasing data availability and the emergence of new analytics methodologies (von Stosch et al., 2014; Sansana et al., 2021).

Managing process changes in CPI requires moving the process from its current operating region A to a new region B. In the first stage, models developed in region A are extrapolated directly to region B. Extrapolation performance is important to obtain reliable estimates of the quality metrics of interest. In the second stage, as the process continues operating in region B and data is collected in this region, model re-training can be performed to account for possible mismatches. Finally, there is a third stage where a large volume data on region B has been collected. Calling attention to the second stage, methodologies that efficiently utilize limited data are preferred. This second task is the focus of the field of transfer learning, allowing one to leverage information obtained in region A to region B.

In this work, we study the application of hybrid modeling for supporting extrapolation and transfer learning, both critical tasks when managing process changes. Hybrid modeling combines physics-based and data-driven models to achieve improved performance. Physics-based models are more challenging to develop and require significant *a priori* knowledge. In contrast, data-driven models are constrained to the data domain where the model is built, and their extrapolation capabilities are limited. Combining both types of modeling mitigates some of their disadvantages while improving predictions (Oliveira, 2004). In this work, we study different configurations of hybrid modeling (e.g., parallel, series) and compare them to benchmarks that include a physics-based model only and data-driven models only. The physics-based model considers simplified reaction kinetics. The set of data-driven methods includes partial least squares (PLS), least absolute shrinkage and selection operator (LASSO), random forest and boosting, support vector regression (SVR), and neural networks (NN).

We consider a simulated case study of biodiesel production (Fernandes et al., 2019), testing two setpoint changes in the biodiesel concentration. Modeling performance in the extrapolation and transfer learning tasks is assessed and guidelines are provided for selecting the most suitable methodologies that show robust performance. Hybrid modeling consistently shows improved results compared to using physics-based or data-driven models only. In particular, parallel hybrid approaches are preferred for the extrapolation task. Regarding the transfer learning task, hybrid modeling also shows advantages, requiring fewer samples than the other benchmarks. Therefore, the application of hybrid modeling to transfer learning, one of the novelties of this work, led to better results than the traditional alternatives.

References

Fernandes NCP, Romanenko A, Reis MS. Mechanistic Modeling and simulation for process data generation. *Industrial & Engineering Chemistry Research*. 2019;58.

Oliveira R. Combining first principles modelling and artificial neural networks: a general framework. *Computers and Chemical Engineering*. 2004;28.

Sansana J, Joswiak MN, Castillo I, Wang Z, Rendall R, Chiang LH, Reis MS. Recent trends on hybrid modeling for Industry 4.0. *Computers and Chemical Engineering*. 2021;151.

von Stosch M, Oliveira R, Peres J, Feyo de Azevedo S. Hybrid semi-parametric modeling in process systems engineering: Past, present and future. *Computers and Chemical Engineering*. 2014;60.

87b The Future of Hazard Analysis and Big Data: With Big Data Comes Big Responsibility

Alyse Keller, Sam Aigen, Elias Quinn, Scott Sanderson, Alison Ballon

Over the course of the history of process safety and security, various efforts have been made to help automate, streamline and improve the Hazard Analysis studies of process hazard analysis (PHA) and security risk assessment (SRA). These improvement efforts include industry standards, computer software, corporate and industry guidelines for the studies as well as other improvements. Recent efforts have been made to improve and automate Hazard Analyses with the introduction of machine learning/ automated intelligence. These improvements range from very simple models and tools such as drop down lists for text selection all the way to very complex methods that use machine learning to partially or fully automate the hazard identification and analysis process.

In this paper, we will discuss various methods that can be used to automate the Hazard Analysis processes for PHA and SRA that will improve and streamline the processes as well as make them more efficient. We will discuss potential benefits to these methods as well as potential drawbacks. We also will discuss how we see the future of conducting Hazard Analyses and the role that humans will play in these studies in the future.

3b The Importance of Carbon Intensity and Compliance to Meet Decarbonization Goals

Kristine Klavers

This presentation will focus on the growing importance of knowing your carbon intensity and compliance requirements wherever you are in the energy supply chain. We will discuss the importance of carbon intensity (CI) and industry standardized measurement and auditing requirements, both for the regulated and voluntary carbon markets. A broad introduction on the direction and drivers of the carbon market; regulated, IRA (Inflation Reduction Act) which will give rise to many capital projects like CCUS, H2, will be provided. The importance of LCA's and CI score on carbon/tax credits, which impact investors and the viability of capital projects will also be discussed.

3a Accelerating Green Hydrogen Via Digitalization - Optimization in Design, Operation and Profitability of a Fully Digitized Green Value Chain

Jens Schmidt

Producing green hydrogen and it's derivatives for solving climate crisis and energy security is a key focus area in todays energy transition and has been significantly accelerated in Europe as a consequence of replacing Russian gas. TES as a green energy company is striving to become the largest eNG (eNG = synthetic natural gas made from green H2 and circular CO2) producer in the world and is therefore currently setting up various import and reconversion hubs as well as production plants around the globe to turn renewable electricity into green molecules.

Digital tools and AI will play a key role in all aspects fo this value chain – design of the plants, operation of the assets in the most efficient way and also maximizing profits by selecting the downstream outlet markets based on dynamic pricing.

Starting with design of plants, the nature of a renewable electricity based H2 production brings variability in feedstock. So the project developer is left with an optimization problem which takes into account variability of power generation, operational constraints of downstream equipment based on turndown ratios and ramp rates, choices between electricity storage and hydrogen storage to decouple units, utilization of capital intense equipment such as electrolyzers vs. efficiency losses of large quantity power storage, etc.

All of these variables are constantly updating and depend on specific location and renewable profile of a location, price levels, etc. Only with proper digital tools and algorithms/optimizers can the project developer find the optimum setup and size of the individual process units for the lowest possible cost of the produced eNG.

During operation digitalization is needed to ensure a non stop monitoring of the assets in operation for maximum efficiency, emission footprint, allow continuous improvement and real time response to events. Here a proper IoT strategy and a link to traditional process control is needed to ensure the assets are operated in the most efficient way, especially given the dynamic ramping required to follow renewable energy supply profiles, etc.

Last not least the green eNG allows to serve different markets from the import hub. It can be supplied as is as green natural gas replacement with carbon recycle from the offtaker, in which case it competes with natural gas plus emission costs. Second it can be reconverted back into green H2 with full capture and return of the CO2 to the upstream hydrogen production for methanation, so the sold product is green H2. Last it can be converted back to green electricity in an oxycombustion power plant, also capturing the entire CO2 and returning it to the upstream hydrogen production for fresh methanation into eNG. Here the product sold is a green uninterrupted electricity PPA, available when no sun or wind is producing green electricity locally, causing high proce fluctuations. TES envisions using AI and optimizers to maximize the margins by forecasting and providing the products in the most valuable markets at any point in time.

133a Learn Data-Driven Engineering with Interactive Modules

John Hedengren

Data engineering with acquisition, transport, curation, and storage of data has emerged as a key role in Industry 4.0. This tutorial session shares new resources for hands-on data-driven engineering to teach important data engineering skills in collection, cleansing, pipelines, storage, and quality assessments. The sequence of learning modules covers introduction to data science in Python & MATLAB, data engineering, machine learning, process automation, optimization, and advanced control. Approximately 5,000 students per day access the freely available learning modules as self-paced tutorials on APMonitor.com or for Professional Development Credit on AIChE Academy. The 11 courses contain Python Jupyter Notebooks or MathWorks Live Scripts to provide interactive learning environments. Solution videos for each exercise guide students through potential solution strategies. One of the courses, Machine Learning for Engineers (https://apmonitor.com/pds), has 36 learning modules and 18 case studies that are specific to engineering practice with classification and regression in computer vision, energy, manufacturing, cybersecurity, and engineering design. This tutorial session gives an overview of the learning modules with a few interactive case studies during the session.

Dr. John Hedengren is a Professor at Brigham Young University in the Chemical Engineering Department. He leads the BYU Process Research and Intelligent Systems Modeling (PRISM) group with a current focus on structured machine learning for optimization of energy systems, unmanned aircraft, and drilling. Prior to BYU he worked in industry for 7 years on nonlinear estimation and predictive control for polymers. His work includes the APMonitor Optimization Suite with extension to Python GEKKO. He led the development of the Arduino-based Temperature Control Lab that is currently used by 70 universities for data science and process control education. His 85 publications span topics of data science, machine learning, smart grid optimization, unmanned aerial systems, and predictive control. He is the Chair of the Education Committee for the IEEE Control Systems Society, Distinguished Lecturer for the Society of Petroleum Engineers (2018-2019), recipient of the 2014 AIChE CAST Division David Himmelblau Award for Innovations in Computer-Based Chemical Engineering, and the 2018 AIChE CAST Division Computing Practice Award. He will share an overview of the BYU PRISM group research in hybrid machine learning methods with Transformer Neural Networks, Transfer learning, and applications in computer vision for worker safety and physics-informed neural networks applied to thermophysical property prediction.

133b Monitoring, Diagnosis and Prognosis of Industrial Processes: Past, Present and Future

Marco Reis

Since its introduction in 1924 by Walter A. Shewhart, Statistical Process Monitoring (SPM) has evolved and adapted to the many technological transformations and several industrial revolutions, while maintaining the core principles essentially unchanged. Currently, there is a rich portfolio of SPM approaches that cover different application scenarios and can detect a wide variety of faults (on sensors, actuators, or in the process). Still, when moving to the second stage of monitoring, i.e., to fault diagnosis, the performance of the available methods is not so dependable, often leading to ambiguous or imprecise conclusions. Furthermore, fault Prognosis is attracting increasing interest in industry, but research is still in the early stages.

In this tutorial session, an overview is provided of several evolution trends that led to different monitoring solutions available today. Focus is given to solutions that are robust and mature, rather than recent proposals yet to be fully scrutinized and tested. Some common misconceptions in the use of SPM are also discussed and clarified, and good practices are shared. The complementary and always important topic of diagnosis is also discussed as well as the increasing interest in fault prognosis.

110a Security in Remote Offshore Wind-Integrated Green Hydrogen Production

Ahmed Elkady, Md Tanjin Amin, Faisal Khan

Green hydrogen is seen as a great potential to decarbonize hard-to-electrify industries and support the growth of renewables by solving the problem of seasonal energy storing due to intermittency. However, hydrogen production through water electrolysis is a novel hazardous process with limited testing. Additionally, several hazards come from extreme operating conditions in offshore locations. Consequently, it's safer to do such operation remotely and with as low as reasonably practical people on board which requires monitoring and control. A great solution to that is to utilize the capabilities of the most recent technologies that accompanied the 4th industrial revolution such as digital twins (DTs), Internet of things (IoT), and cyber-physical systems (CPSs). Such relatively new concepts raise new safety and security challenges.

The current work is on the aspect of analyzing cyber security and safety challenges that arise from remote offshore green hydrogen production. It analyzes the literature to investigate opportunities and challenges presented by the application of CPSs in hydrogen production facilities. The analysis attempt to answer following questions: (i) What are the technological enablers for offshore water electrolysis process safety? (ii) What are the characteristics of industry 4.0 technologies that are likely to cause safety and security issues? (iii) What are the safety and security issues introduced by industry 4.0? How can these issues be addressed? (iv) What are other technical challenges related to safety and security?

It is observed that technological enablers include:

- Digital twins (DTs): model multiple designs and scenarios, including variables such as weather, to optimize the process, maximize return on investment and minimize risk.
- Internet of things (IoT) and cyber-physical systems (CPSs): monitor process using key performance indicators (KPI) such as energy consumption and production rates which facilitates rabid anomaly detection. It can also help in safety compliance management and prevent violation acts and offer real time alert management systems. Finally, with remote control capabilities, IoT can help in controlling safety levels during normal and emergency conditions

CPSs have characteristics that introduces safety and security challenges. This includes complexity, heterogeneity, interdependency between cyber and physical processes. Other characteristics include fragmentation, autonomous reconfiguration decisions, and other constrains like physical and resource. CPS autonomous reconfiguration decisions and compensatory actions may cause fault masking effects. There are also security-related issues including integrity and availability challenges in the presence of cyber-physical threats with unknown external disturbances. More challenges include resilience assessment, and human factors. This paper provides a comprehensive risk assessment and management that should proactively identify these potential technical problems so that appropriate actions can be taken to reduce or eliminate the probability and/or impact of these problems.

110b A Novel Zone-Based Machine Learning Approach for the Prediction of the Performance of Industrial Flares

Jian Fang, Helen Lou, Sidney Lin

Industrial flares are used to burn off unwanted gas during operation. If not combusted completely, intermediate products or incomplete combustion products are formed, and they will cause significant environmental and health issues. The EPA Refinery Sector Rule emphasizes smokeless flaring with combustion efficiency (CE) $\geq$ 96.5% and destruction and removal efficiency (DRE) $\geq$ 98% for all types of flares in the refineries. In this research, a novel zone-based modeling approach was developed for predicting CE and Opacity of steam assist flares. Flare CE data were partitioned into two zones based on the partition of the carbon and hydrogen atomic ratio (CHR), then random forest (RF) and CatBoost algorithms were used to develop CE predictive models respectively. This CHR-based zone partition has a clear implication in engineering. It was also found out that no zone division for flare Opacity prediction is needed, and both RF and CatBoost algorithms generated good prediction results. All the models match extremely well with all the original experimental data. These predictive models under the same zone-partition use either RF or CatBoost algorithm can both give superior prediction accuracy. This demonstrates the simplicity, general applicability, and high reliability of the zone-based ML approach.

110c Joint Angle Calculations Using Motion Capture and Deep Learning Pose Estimation for Safety Applications

Luke VanKeersbilck, McGyver Clark, Iain Hunter, John Hedengren

Marker based motion capture is currently the most accurate method of measuring human kinematics, but it is limited to tracking objects with reflective markers making it unsuitable for human pose tracking applications in most work environments. The purpose of this study is to quantify the accuracy of pose estimation from a monocular electro-optical sensor with deep learning to infer segment end points and pose estimation. An accuracy assessment is performed with a Vicon Nexus and an iPhone, both running at 240 Hz. Visual 3D joint angles are computed from the marker data. The iPhone view was placed perpendicular to the sagittal plane. Deep learning algorithms produced 3D pose information that is translated into hip, knee, and ankle sagittal plane joint angles. Pearson r correlations compared MediaPipe joint angles through running, lifting, ladder climbing, and step motion capture data. On the running data, markerless methods showed correlation values compared with Visual 3D of hip (0.968), knee (0.983), and ankle (0.928). The markerless methods have limitations on predicting maximum flexion and extension angles. Although the correlation values are high, in practice these differences in maximum range of motion may impact any future interpretation of data. Care should be taken when utilizing extreme joint angles when using deep learning algorithms. Although at this point the open-source methods are not as accurate as marker-based motion capture, they could enable the collection of data from workers in a manufacturing environment. Given the ease of data collection, this could facilitate safety tracking applications in work areas that require safe human pose or movements. Real-time pose estimation is demonstrated with a classifier to assess office desk ergonomics.

110d How to Implement AI in a Data Lean Environment to Deliver Business Value.

Forogh Askari

The promise of AI used to be contingent on the need for big volumes of data. However, AI is not as data-hungry as once thought. New techniques in industrial AI and expert practices in data pre-processing combined with minimum good quality data are proven to deliver true benefits. In this presentation, Canvass AI will discuss the key data requirements that engineers need to get started with AI to derive real value. We'll explore how to assess data sufficiency, including quantity, quality, and descriptiveness. In addition, we'll address how to integrate experience into the data, particularly as it regards the institutionalization of knowledge and advanced enablement.

This presentation includes real-world case studies of how chemical engineers are extracting value from their existing data to simplify their decision-making and fast-track their day-to-day problem-solving. The presentation will include a deep dive into one of the case studies to share how data and model building steps, and specialized Industrial AI capabilities can analyze process dynamics and narrow down the problem size to create AI models that are highly accurate and reliable.

110f How the Start-Small Strategy Is Helping Chemical Companies Generate Millions in Savings and Drive a Sustainable Future

Forogh Askari

An estimated 60% - 80% of organizations fail to report significant business gains from AI. There are many reasons why AI fails to deliver on its promise, and the main reason is that companies start with problems that have too many organizational complexities, and that take too long to realize value. Instead, by starting small and targeting specific AI solutions that produce immediate business value, organizations can effectively take the first steps in their AI journeys while preparing their workforce for the more complex applications that unlock the full value of AI.

As part of this presentation, we'll explore a case study where engineers at a chemical company did just that. They started with a smaller opportunity, addressed it to provide immediate benefits and build trust within the organization, and then applied those learnings to adjacent challenges and assets to improve the overall asset reliability, safety, and productivity at their production facility. By making AI solutions practical for engineers and operators from the beginning, the workforce was able to scale AI across over 20 assets and processes, resulting in nearly \$20M/year in benefits. The presentation will include the implementation process learned from over hundred implementations of AI projects. The systematic process covers all the essential parts of successful AI implementations. Don't make the mistakes of others, instead, learn how to start on the right foot and fast-track your AI journey to success.

TITLE:

Enabling Autonomous Operations

AUTHORS:

- Oleg Mikhailov, CEO, Xplorobot oleg@xplorobot.com
- Richard Clarke, CTO, Xplorobot richard@xplorobot.com

ABSTRACT:

Each use case for autonomous operations is different and there is currently no one-size-fits-all solution available on the market. We will present our experience with multiple robots, drones and sensors at various locations both onshore and offshore, covering installation, data acquisition and processing, as well as the key challenges and results.

There are numerous factors affecting the choice of robot for data acquisition. For example, many areas require equipment which is certified as intrinsically safe due to potentially explosive gases. There are very few such robots, and if additional sensors are required then recertification is necessary, which can be a long process. Wheeled robots typically have a longer range than legged robots, and although they can avoid small obstacles, they are not able to step over broad obstacles such as hoses. In some cases, the terrain may be such that a hand-held sensor package is preferred, or the inspections are infrequent and investing in automated data acquisition does not make economic sense.

The selection of sensors is also very important and depends on the particular use case. Each high-end sensor may represent a significant investment, so we need to ensure that they are fit for purpose, and we do not carry redundant sensors. Note that our "mobile sensor" approach has several advantages over an IoT solutions, particular at large sites where the large number of fixed sensors would be cost prohibitive. It is flexible: if a site is modified, the robots can be given a new route to monitor. IoT sensors require permits and installation, and as a fixed part of the site, themselves need inspecting. The mobile approach is also future proof as sensors can be easily upgraded as technology advances.

The key to a successful deployment however is the integration of all this hardware through well designed software, from data acquisition, though an automated data processing pipeline to visualization of the results that allows engineers to make decisions quickly. Rather than focus on reading gauges, as is commonly the case, we prefer to acquire dense data about the entire site. A typical inspection can thus generate ten to twenty gigabytes of data per hour from multiple different types of sensors and creates significant data management challenges.

After an initial site visit and 3D model building stage, we can process the data "at Edge" and distil the dense data down to easily interpretable heat maps, gas concentration maps, etc, in quasi real time. This process leads to a data compression of up to a thousand, allowing the results to be transmitted over low bandwidth networks, such as 4G, to the remote engineers for analysis. The use of 3D models facilitates time lapse analysis even when the data is acquired from different locations due to obstacles on the inspection route, for example. This allows to generate alerts when an inspection result is outside of normal conditions. With longer historical data, we can use repeat visual data to detect areas with higher corrosion rates, which normally go unnoticed by humans.

Finally, it is obvious that regular inspections followed by this processing build an incredibly rich and organized database that can be used for predictive analytics that will generate insights and solutions that we are only just starting to imagine.

KEYWORDS:

- Anomaly detection
- Artificial intelligence
- Automation
- Autonomous operations
- Data analytics
- Digital Transformation
- Digital twin
- Governance & Regulation
- Health, Safety, Environment
- Inspection
- Integrity Management
- Maintenance
- Methane Emissions
- Optimisation
- Robotics
- Sustainability

139c Enabling Autonomous Operations

Richard Clarke, Oleg Mikhailov

Each use case for autonomous operations is different and there is currently no one-size-fits-all solution available on the market. We will present our experience with multiple robots, drones and sensors at various locations both onshore and offshore, covering installation, data acquisition and processing, as well as the key challenges and results.

There are numerous factors affecting the choice of robot for data acquisition. For example, many areas require equipment which is certified as intrinsically safe due to potentially explosive gases. There are very few such robots, and if additional sensors are required then recertification is necessary, which can be a long process. Wheeled robots typically have a longer range than legged robots, and although they can avoid small obstacles, they are not able to step over broad obstacles such as hoses. In some cases, the terrain may be such that a hand-held sensor package is preferred, or the inspections are infrequent and investing in automated data acquisition does not make economic sense.

The selection of sensors is also very important and depends on the particular use case. Each high-end sensor may represent a significant investment, so we need to ensure that they are fit for purpose, and we do not carry redundant sensors. Note that our "mobile sensor" approach has several advantages over an IoT solutions, particular at large sites where the large number of fixed sensors would be cost prohibitive. It is flexible: if a site is modified, the robots can be given a new route to monitor. IoT sensors require permits and installation, and as a fixed part of the site, themselves need inspecting. The mobile approach is also future proof as sensors can be easily upgraded as technology advances.

The key to a successful deployment however is the integration of all this hardware through well designed software, from data acquisition, though an automated data processing pipeline to visualization of the results that allows engineers to make decisions quickly. Rather than focus on reading gauges, as is commonly the case, we prefer to acquire dense data about the entire site. A typical inspection can thus generate ten to twenty gigabytes of data per hour from multiple different types of sensors and creates significant data management challenges.

After an initial site visit and 3D model building stage, we can process the data "at Edge" and distil the dense data down to easily interpretable heat maps, gas concentration maps, etc, in quasi real time. This process leads to a data compression of up to a thousand, allowing the results to be transmitted over low bandwidth networks, such as 4G, to the remote engineers for analysis. The use of 3D models facilitates time lapse analysis even when the data is acquired from different locations due to obstacles on the inspection route, for example. This allows to generate alerts when an inspection result is outside of normal conditions. With longer historical data, we can use repeat visual data to detect areas with higher corrosion rates, which normally go unnoticed by humans.

Finally, it is obvious that regular inspections followed by this processing build an incredibly rich and organized database that can be used for predictive analytics that will generate insights and solutions that we are only just starting to imagine.

117a Trends in on-Line Gas Chromatography and Unmet Needs

Eric G. Schmidt, Wayne Kubala, Rod Spitler

On-line Gas Chromatography (GC) has been a part of the chemical industry since the mid 1950s. Many technological advances have been made in the industry since the first GC including improvements in sample conditioning systems, injection valves, columns, detectors, and finally data acquisition systems. As technology continues to develop, there is considerable interest in the chemical industry to install on-line analyzers or sensors near the sample point to both decrease sample transport time as well as eliminate the capital expense of installing a large shelter and the long term shelter maintenance.

This presentation will focus on recent technical improvements in on-line GC as well as discussion of the future unmet needs for process on-line gas chromatography, including using a shelterless gas chromatograph for analysis in chemical plants. In addition, a broader discussion of sample system design features including safety, sample integrity, reliability, maintainability, and cost will be included. An example installation of an on-line hydrocarbon analysis with a field mounted gas chromatograph will also be discussed. We will compare and contrast the specific technical merits of traditional gas chromatographs installed in climate controlled shelters and shelterless gas chromatographs.

117c Use of TDLAS in Fired Assets to Aid in Decarbonization Efforts

Ryan McSherry

Global efforts are well underway in a variety of industries to reduce greenhouse gas emissions. In order to achieve goals set by the 2015 Paris Agreement and 2018 UN IPCC revision, global emissions must be reduced 45% by 2030, and 100% by 2015 to bring atmospheric warming levels within a safe region. Industrial fossil fuel combustion processes generate about 16% of total U.S. carbon dioxide emissions (13% of total U.S. greenhouse gas emissions), and combustion of fossil fuels generates 31% of total U.S. carbon dioxide emissions (24% of total U.S. greenhouse gas emissions) as of 2020. Combined, combustion assets account for about 47% of total carbon dioxide emissions in the US, and 37% of total U.S. greenhouse gas emissions. Naturally, there's a heavy focus in the U.S. on implementing deep decarbonization strategies to support efforts for net-zero greenhouse gas emissions by 2050. This topic will focus on how developments in TDLAS technology can support advanced control efforts for fired assets to safely reduce carbon dioxide emissions via less fuel usage and increases in efficiency.

117b Non-Contact Gas Spectroscopy Exploting Tuning Fork-Based Light Detectors

Angelo Sampaolo, Pietro Patimisco, Andrea Zifarelli, Giansergio Menduni, Marilena Giglio, Hongpeng Wu, Lei Dong, Vincenzo Spagnolo

In the past decade, the rapid development of infrared laser technology has led to an increasing demand for photodetectors with high sensitivity and a wide operative spectral range suitable for spectroscopic applications. In this work, we report on the study of light-induced thermo-elastic effects occurring in quartz tuning forks (QTFs) when exploited as light detectors in Tunable Diode Laser Absorption Spectroscopy (TDLAS) spectrometers for non-contact gas sensing [1]. The induced photothermal processes and the temperature distribution following the absorption of laser beam upon the crystal quartz was studied by using finite-element-analysis with COMSOL Multiphysics. The electromagnetic energy release and the induced thermal distribution were related to the absorbance curve of the quartz crystal, shown in Fig.1a. In the spectral region with high absorption, the radiation travels few tens of micrometers in the quartz crystal, while in the spectral region with low absorption, the radiation is trapped at the interface between the chromium film and the highly reflective gold layer [2].

Figure 1: (a) Transmittance spectrum of the QTF quartz measured using a Fourier transform spectrometer (continuous line). A 750 μm-thick quartz slab was employed for the measurement. (b) Responsivity as a function of the wavelength for three models of photovoltaic Vigo detectors, namely, PVI-4 (red line), PVI-6 (blue line), and PVI-10.6 (green line) HgCdTe, and for Thorlabs PDA10CF InGaAs Amplified Photodetector (black line), as provided by the device manufacturers. Responsivity of the QTF (black squares) employed as a photodetector.

The spectral response of the QTF-based photodetector was investigated by using a custom QTF with a resonance frequency of 9.78 kHz and quality factor of 11500 at atmospheric pressure. Five interchangeable laser sources operating at different wavelengths from 1.6 up to 10.35 μm were employed within TDLAS sensors. A spectrally flat responsivity of 2.2 kV/W was demonstrated (see Fig.1b), corresponding to a noise-equivalent power of 1.5 $nW/Hz^{1/2}$, without employing any thermoelectrical cooling system. This tuning fork-based sensing configuration pave the way to low cost, modular spectrometers capable to perform non-contact analysis of continuous flow gas streams with highly variable composition and with a sensitivity in the part-per-billion range.

References:

[1] S. Dello Russo et al., Opt. Express 28, 19074-19084 (2020).

[2] T. Wei et al., Appl. Phys. Rev. 8, 041409 (2021).

150b The Next Epoch of Model Predictive Control: Exploiting Machine Learning Methods for Approximation and Design

Joel Paulson

Model predictive control (MPC) is an optimization-based predictive control strategy that has become very popular in the last few decades due to its broad capabilities that include accounting for systems with multivariable and interacting dynamics, nonlinearities, and constraints. Furthermore, its versatility in terms of its ability to provide robustness guarantees (by design) as well as consideration of economically oriented control objectives, MPC is increasingly being applied in many emerging systems including chemical process systems, energy systems, robotics, unmanned vehicles, and biomedical systems, to name a few. Yet, many MPC applications face important challenges related to the difficulty of modeling complex systems and the need for more advanced MPC strategies to provide provably safe and robust performance with low online computational and memory requirements. The proliferation of machine learning (ML) and artificial intelligence (AI), combined with the availability of increased computational and sensing capabilities in modern control systems, has led to a rapidly growing interest in the use of learning-based methods in MPC. While most research in this area has focused on the construction of the internal prediction model, the opportunities for ML/AI extend far beyond model improvement and adaptation. In this tutorial session, we will provide an overview of recent efforts for fusing ML/AI and MPC as well as discuss potential future research opportunities in this rapidly evolving field.

140a Benefits of Designing Product Development Experiments with AI

Kristin Wallace, Marlene Cardin, Alexander Nguyen

Gone are the days where a small group of domain experts plan single factor or simple mixture DOEs, and then patiently wait for these experiments to be scheduled into busy laboratories or pilot plants. R&D teams across many industries are embracing AI-powered tools to transform workflows and accelerate innovation.

This presentation will demonstrate the benefits of designing experiments with FormuSense. ProSensus developed FormuSense to be a streamlined implementation of our proven product development framework, which is centered around multivariate analysis.

A foods dataset (of baked goods) will examined in FormuSense to discuss the intuitive visualizations and features available that help formulators rapidly understand key correlations among and between ingredient properties, formulation ratios, process conditions, and final quality. The no-code platform will be used to suggest a small number of useful experiments to explore new or underrepresented regions of the design space.

140b Advanced Analytics Role in Technical Workforce Transformation

Rupesh Parbhoo, Allison Buenemann

Data has become the primary driver of ROI in industries that previously saw capital investment in plant equipment as the main means to debottleneck processes. A popular misconception is that in order to effectively analyze data or leverage machine learning techniques, a data science background is needed. While a knowledge of data science is useful, a knowledge of the plant – how it's designed, how it's operated, how the process has adapted over time – is imperative. Advanced analytics software Seeq is putting the power of data science and machine learning in the hands of those who know the process best. Wrapping advanced algorithms in point-and-click experiences lowers the activation energy for solving previously unsolvable problems enabling organizations' existing workforce to capture business value without capital investment. This approach has led chemicals and other process industry companies to millions in reliability, productivity, quality, and sustainability improvements. In this talk, we'll share best practices on software rollouts that transform the technical workforce and a use case identification methodology that scales with the organizations' increasing analytics capabilities.

140c Moving Beyond Manufacturing Analytics to Knowledge

Peter Guilfoyle

Manufacturing analytics providers have long been focused on helping customers improve operations and processes through the application of univariate and multivariate SPC as well as other advanced analytics methodologies.

Industry 4.0 and Digital Transformation initiatives were important drivers of the uptake and robust usage of manufacturing analytics to deliver performance improvements. Manufacturers were able to transition from the usual react & respond production paradigm to predict & prevent – predictively identifying and pro-actively preventing issues through the use of real-time manufacturing analytics *before* problems negatively impact the environment and bottom line.

Over the years, many vendors have improved their customers' abilities to unlock more value from their existing data through sometimes novel (and often not very scalable) solutions. During that time, analytics have become (essential) table stakes in an ever-increasingly competitive manufacturing market.

From inbound raw materials to outbound finished goods and all stops in between, manufacturers are applying analytics to influence positive process outcomes, end-product quality, and customer satisfaction. They are driving immediate course corrections necessary to keep processes in control and product within specification. They are making analytics-based decisions to improve production efficiencies, increase supplier performance, and decrease waste.

What manufacturers are failing to do, however, is deliver lasting (and repeatable) value to their organizations with manufacturing analytics.

The next step-change evolution of manufacturing analytics is capturing the intellectual property that results from their use. Analytics-based knowledge spans the identification of early-warning signals, the codification of institutional knowledge and the amplification of process knowledge for quick and decisive action.

Attracting, retaining, and empowering workers should also be a key focus of any analytics technology investment. By capturing the knowledge that results from the application of manufacturing analytics (from data through result), a company builds a foundation to not only solve process issues faster but as importantly addresses three of manufacturing's biggest human capital challenges[1] – the loss of institutional process knowledge through retirement; ever-increasing skills gaps between long-term employees and others; and time-to-productivity for new hires.

The session will feature real-life use cases and a live demonstration of how manufacturers are delivering on the promise of Industry 4.0 and Digital Transformation as they:

- Leverage existing process-related data and applying a wide variety of manufacturing analytics methodologies (e.g., univariate and multivariate SPC; machine learning, etc.), rules, formulas, and models to proactively identify and alert on potential process issues and problems.
- Connect specific actions and quantified best practices to each analytics alert and delivers those directly to users.
- Preserve process knowledge by retaining all relevant event information – data, analytics, actions, and results – for use anywhere across the enterprise as well as delivering better work processes for continuous process improvement and employee onboarding.

[1] Deloitte 2022 Manufacturing Industry Outlook

104a Multi-Fidelity Surrogate Models to Generate Low-Fidelity Data for Data-Driven Branch-and-Bound Optimization

Suryateja Ravutla, Fani Boukouvala

High-Fidelity (HF) Simulations are essential in quantitative analysis and decision making in many engineering fields. However, some of the challenges when optimizing with embedded HF simulations are: (a) the lack of objective function and constraint equations and derivatives, and (b) high-computational cost (van de Berg et al., 2022). To optimize such systems, black-box optimization methods are used. We recently proposed a data-driven equivalent of spatial branch-and-bound algorithm (DDSBB) based on the concept of constructing underestimators to simulations from the HF data (Zhai & Boukouvala, 2022). The algorithm uses a linear programming formulation to fit quadratic underestimators that satisfy constraints that enforce their validity based on collected samples and estimated Lipschitz constants. This algorithm has shown promising results in identifying globally optimal solutions with less dependence on the initialization, and provides bounds on the identified optimal solutions (i.e., worst-best case optimal solutions) at any point of the optimization process. However, the algorithm requires a large amount of samples to converge, especially with increasing dimensionality.

One way to reduce the required HF sampling cost is to utilize the HF data to construct Low-fidelity (LF) surrogates, and generate LF data to improve the underestimators. In our prior work, we have shown that utilizing LF predictions lead to improved performance (Zhai & Boukouvala, 2022). Recent advances in hybrid surrogate modeling and physics-informed machine learning (Bradley et al., 2022), provide new opportunities for further improving this framework and we intend to explore many such techniques in this paper. We start by quantifying the effect of different LF hybrid surrogates on performance of our algorithm, by studying a large set of non-linear benchmarking problems. While this approach provides an advantage as is, we will devise techniques to further improve the LF data accuracy by utilizing the concept of multi-fidelity surrogate models (MFSMs). We will quantify the improvements in black-box optimization, when taking advantage of the feature of MFSMs to "learn better", by exploring correlations between the LF and HF data (Guo, Manzoni, Amendt, Conti, & Hesthaven, 2022). The performance and implementation of this work will be shown in the presentation through our existing python package (PyDDSBB).

References

Bradley, W., Kim, J., Kilwein, Z., Blakely, L., Eydenberg, M., Jalvin, J., . . . Boukouvala, F. (2022). Perspectives on the integration between first-principles and data-driven modeling. *Computers & Chemical Engineering, 166*, 107898. doi:https://doi.org/10.1016/j.compchemeng.2022.107898

Guo, M., Manzoni, A., Amendt, M., Conti, P., & Hesthaven, J. S. (2022). Multi-fidelity regression using artificial neural networks: efficient approximation of parameter-dependent output quantities. *Computer methods in applied mechanics and engineering, 389*, 114378-114378.

van de Berg, D., Savage, T., Petsagkourakis, P., Zhang, D., Shah, N., & del Rio-Chanona, E. A. (2022). Data-driven optimization for process systems engineering applications. *Chemical Engineering Science, 248*, 117135.

Zhai, J., & Boukouvala, F. (2022). Data-driven spatial branch-and-bound algorithms for box-constrained simulation-based optimization. *Journal of Global Optimization, 82*(1), 21-50.

104b Considerations for Product Development Model Formulations in Constrained Optimization Problems

Alexander Nguyen, Marlene Cardin, Kristin Wallace

For manufacturers, the continuous development of new products is necessary to remain competitive, whether it is to reduce cost, utilize alternative materials, or to meet new regulations. A design-of-experiments approach may not be the most effective tool to explore formulation options, especially for products that contain many ingredients and properties. Instead, to capitalize on data collected during product development, multivariate analysis (MVA) can be used to build a model to predict final product quality based on ingredient mixture properties, along with other fundamental knowledge.

In tandem with these statistical models, there are factors to consider that may not be captured in the model correlation structure, such as the importance of cost and quality, the introduction of new ingredients, and constraints on ingredient ratios, classes, and mixture properties. In order to explore solutions of this multifaceted problem, a projection onto latent structures (PLS) model is trained and embedded in a nonlinear constrained optimization problem.

This presentation will provide a walkthrough of the embedded optimization problem, how modeling decisions are considered in the formulation, and practical solution approaches.

104c Achieving Targeted Product Mix Using Web-Based Optimization Tool for Chemical Plants with Multiple Products

Zubin Kuvadia, Xinjie Tong, David Koscielniak, Shachit Iyer, Jianping Zeng

Many chemical manufacturing plants are designed to co-produce multiple products which may differ with respect to factors such as chain length or degree of substitution. Market demand will determine the actual production requirement for each individual product and may vary widely over the course of a month, year, or longer. Multiple controllable conditions can be used to achieve the desired production mix and suppress unwanted by-products. The sheer number of variables and process constraints make meeting multiple production requirements a challenging task. A tool that guides decision-making using near-real time optimization control can maximize the overall profit of the plant. Moreover, coordination with other plants is also very important to meet the strategy of the business. Currently, this multi-dimensional control process highly relies on operators' experience aided by numerical calculation using tools like Excel which support what-if scenarios. Thus, to find the optimal operation conditions for each plant in real time is inefficient and may not yield the optimal result. The goal of this research work is to develop an inverse optimization tool, which uses desired production mix and other decision variables to determine the optimal control conditions, to help operators make the decision more efficiently and reliably. This web-based optimization tool will combine existing numerical models and safety constraints. This tool has helped production plants to increase production rate and achieve additional margin. Ongoing change management and improvement efforts for the tool will also be briefly highlighted.

71a Data-Driven Transformations for a Sustainable Enterprise

Stephen Reynolds

There is a tremendous amount of data accessible in our operating units and businesses. From early design and modeling to project execution, from startup to optimization, we are constantly compiling more and more information on the processes we manage. Science says we don't use most of it. In this presentation, we'll look at the lifecycle of a chemical plant and the people that work there and identify the data sources that we can use to enable meaningful transformation. At the heart of this transformation lies the digital twin, connecting the data sources, empowering the connected worker as they translate that data into information and ultimately into action. Data is a modern utility stream that drives the transformation into a more efficient, sustainable enterprise.

71b Impact of HTML5 and Web-Based Solutions for Frontline Plant Workers.

Iiro Esko

The purpose of this abstract is to share insight on how a move from a client-server architecture to web-based architecture impacts frontline plant workers in process industries.

Mobile access to operational technology (OT) solutions such as a process control system, manufacturing execution system, engineering databases such as 2D and 3D digital twins, maintenance databases, process simulation, and data visualization through HTML5 interfaces can be significantly faster than a traditional client-server approach.

This abstract will discuss business benefits to organizations, work related benefits for frontline workers like technicians, engineers, and managers, as well as recent developments in automation, project, maintenance, and production engineering spaces. The author will share individual application examples in engineering, simulation, process control, maintenance, and plant up-keep, as well as a full stack application example available for visits by industry professionals at a US Government funded innovation institute located in Chicaco, IL.

This abstract is recommended to technology leaders recommending solutions for new projects and plants, for technology experts looking for solutions for existing plants, and for young professionals looking to gain knowledge of topics related to mobile workers, augmented reality, remote support, use of digital twin data, process control, process simulation, HTML5 interfaces, data visualization, and digital enterprise concepts.

The author promotes a complete stack of digitalization and automation technologies towards chemical process industry and recently acted as a project manager for a demonstration skid focusing on a mobile worker experience based on digital twins.

71c Safety Digitalization

Abdulrahman Alamer, Wassam Alqahtani

Digital transformation is expected to change how we live, work and communicate. The term is used to describe the current automation trend and data exchange. It typically describes the integration of several existing and developing technologies including mobility, wearables, Internet of Things (IoTs), and Augmented and Virtual Reality applications. Safety, like many other fields, stands to benefit greatly from this transformation. Ease of access to information, increased input of meaningful data, and proactive and predictive safety are but a few examples of these benefits. The presentation would provide samples of the successful safety digitalization efforts within Aramco including mobility, predictive safety analytics and Virtual Reality Safety Training. Aramco's SafeLife, a 'One-Stop Shop' for safety applications, is a digital platform that enables Aramco organizations and contractors to manage their safety while providing oversight, monitoring, and analytical capabilities. SafeLife encompasses a wide spectrum of digitalized safety processes and applications. These include incident reporting, inspection management, dashboards and reports, Risk Management, Competency and Training, and Leadership Engagement. In the field of training, Aramco is overhauling its corporate training to introduce active learning curriculums through various AR/VR applications. Safety training, is one of the areas where major improvements have been achieved. One of the challenges that always face safety training, is the difficulty in conducting hands on training. More often than not, it is unsafe or even dangerous to expose trainees to real working environments for the purposes of training. Hence why bringing the environment in the form of Immersive Virtual Reality enables us to overcome this challenge. IVR has been proven to enhance understanding, improve knowledge retention, and provides realistic scenarios tailored towards the audience.

97a Cybersecurity Risk Assessment Strategies in Industrial Control Systems

Timothy Gale

Industrial Control Systems (ICS) have become more vulnerable with the increasing use of commercially available off the shelf (COTS) technology, enterprise integration and remote access requirements. The cybersecurity risks associated with these vulnerabilities can be difficult to capture and present as they include traditional process safety consequences and information technology causes. This presentation will look at some strategies to conduct Cyber Risk Assessment and some best practices for success.

53a Optimizing Petrochemical Manufacturing Processes with Deep Learning

Roberto Linares

The highly competitive environment of petrochemical production requires a high utilization and optimization of the processes to produce chemicals with the least amount of energy and cost.

The petrochemical industry has invested in control systems, advanced process control and real time optimization applications to optimize their process and operations. However, the current technologies do not optimally address changes in feed composition and other effects related to asset utilization. Artificial intelligence (AI) and machine learning (ML) powered by cloud-based computing offers a complementary add-on to increase efficiency and support higher yields and margins by operating closer to product specification at the lowest energy intensity. AI can optimize the existing automation platforms and applications by providing real time predictions and prescriptions for operators and plant engineers.

This presentation addresses the development and deployment of AI-based advanced process control and real time optimization applications designed to decrease emissions, reduce energy consumption, and improve overall efficiency of petrochemical plants.

53b *An Adaptive Framework for State-Based Prediction of Rotating Equipment Failures*

Maxwell Toothman, Birgit Braun, Scott J. Bury, James Moyne, Dawn Tilbury, Kira Barton

An Adaptive Framework for State-Based Prediction of Rotating Equipment Failures

Motivation and Problem Statement

Rising global demand is motivating the manufacturing industry to implement maintenance strategies that minimize operational downtime. The field of prognostics and health management (PHM) research seeks to develop methods and frameworks that minimize downtime by avoiding both equipment failure and unnecessary preventative maintenance [1]. A key aspect of PHM research is the development of predictive maintenance (PdM) models, which detect ongoing equipment degradation and predict system failures. Because rotating equipment (such as pumps, turbines, and compressors) is widely used in process manufacturing operations, much of the recent PdM research has focused on this class of equipment.

Two limitations of existing PdM models have hindered their efficacy when applied to industrial rotating equipment. The first is a reliance on instantaneous (snapshot) analysis, which involves using only the most recent measurements from a system to detect degradation and predict failure. PdM models that take this approach are often unable to recognize the early stages of degradation in rotating equipment, which can be characterized by gradual trends in machine signals. An alternative to snapshot analysis is state-based analysis, where a PdM model uses histories of recent system measurements to classify the current state of equipment health and generate predictions about future health states. The second limitation of existing PdM models is treating system health models as static entities. While the shape and direction of degradation generally remains consistent across different machines, the value of healthy signal baselines and the speed of degradation can vary significantly, even throughout the life of a single machine.

PdM models that do not detect and adapt to variances in degradation behavior can generate inaccurate modeling outputs and quickly become obsolete.

This work addresses these limitations by proposing an adaptive framework for

predicting failures in industrial rotating equipment with state-based PdM models. The work here is an extension of the methods developed in [2] and [3], which use static, state-based models to make repair quality assessments, and the adaptive modeling framework proposed in [4], which is specific to rolling element bearing degradation. The proposed framework relies on equipment and reliability experts to specify a set of degradation modes to which a system is susceptible, along with templates for models describing the general behavior of machine signals as a function of time. Soliciting this input increases the resources necessary to implement the framework but allows it to be applied across a wide range of systems and degradation modes and results in modeling outputs that can easily be interpreted by end-users.

Approach

1. Modeling Architecture

This framework relies on equipment and reliability experts to define a set of degradation modes to which a machine is susceptible, and a predictive model class for each degradation mode. As in object-oriented programming, predictive model classes act as a template for creating model instances that describe the degradation behavior of one or more features derived from machine signals. Predictive model classes consist of a set of fixed parameters, with values that remain constant across different machines and failure events, as well as a set of variable parameters, with values that are allowed to vary across machines and failure events. Additionally, a predictive model class must include a method that maps the duration of degradation to a probability density function describing the value of one or more system features. For example, a predictive model class may describe a linear increase in a vibration feature () over time () using the function , where the slope parameter () is fixed, the variance of a Gaussian random variable () is fixed, and the offset parameter () is variable. Finally, a predictive model class must include a method to re-estimate the local parameters of a model based on a training dataset.

During online monitoring, a system health statechart describes the degradation modes that a system may experience. These statecharts consist of a state and a set of states that are each associated with a predictive model, instantiated from a pre-existing model class, describing the expected degradation behavior in that state. The state is associated with a model instance that describes steady-state behavior in machine features, which is referred to as a stationary model. Once a system's potential degradation modes and associated predictive model classes have been defined, the adaptive methodology described in the following section can be implemented to perform state estimation and failure prediction.

1. Adaptive Failure Prediction Methodology

This methodology defines four computational stages that are used to monitor the health of a system during operation and adapt existing modeling resources when system measurements deviate from expected behavior. The first stage of the methodology, **routine monitoring**, involves two processes: state estimation and anomaly detection, that are repeated whenever new measurements are collected. The goal of state estimation is to determine the recent state history of a system based on the current structure of the system health statechart. This is implemented by adapting the Viterbi algorithm, a common method for estimating the state histories of systems represented by hidden Markov models (HMMs) [5], to be compatible with the time-varying feature distributions provided by predictive models. State estimation results provide estimates about the current degradation mode of a system based on recent measurement histories along with an estimate of the degradation onset time, defined as the time at which a system transitions from the state to the most-likely state. Well-established Monte Carlo methods can then be used to estimate a time-to-failure distribution for the system based on the predictive model associated with the most-likely degradation mode [6].

Anomaly detection seeks to identify when system measurements start to deviate from the behavior described by the state's stationary model or any of the predictive models included in a system's health statechart. To achieve this, the system health statechart is again considered to be an HMM () and a Markov chain Monte Carlo sampling method is used to generate samples from the observation probability distribution of the HMM, . The probability of the most recent observation is then compared with these samples to determine whether the observation is anomalous or expected. When several consecutive anomalous observations are detected, the subsequent stages of the methodology provide mechanisms to adapt the predictive signal models and system health statechart in response.

The second stage of the methodology, **model localization**, involves re-estimating each model's variable parameters based on recent system measurements. Anomaly detection is then repeated with the updated models and, if the anomalies are not resolved, an **automated statechart update** stage is triggered. This process provides a mechanism to modify the degradation stages included in a system's health statechart. The PCA similarity factor [7] between recent measurements and the entire set of available model classes is computed to quantify whether any models capture the general direction of degradation. If any model classes demonstrate a sufficiently high similarity factor, the current statechart is augmented with a state for each of these model classes, and anomaly detection is repeated. In the case that anomalies are still not resolved, records of recent measurements and the results of the PCA similarity analysis are reported to system experts to inform the creation of a new model class to describe this unexpected degradation behavior as part of the final stage, **SME-informed statechart update**.

Concluding Remarks

The framework developed here enables state-based failure prediction with degradation models defined by experts and provides a methodology for adapting these models when anomalous degradation behavior emerges. These capabilities provide transparent modeling results to inform maintenance decisions and facilitates the deployment of robust PdM strategies in industry. Case studies in [3] and [4] demonstrate the value of an adaptive, state-based modeling approach, and the full presentation for this project contains a case study implementing the framework described here.

References

[1] J. Lee, F. Wu, W. Zhao, M. Ghaffari, L. Liao, and D. Siegel, "Prognostics and health management design for rotary machinery systems reviews, methodology, and applications," Me-

chanical Systems and Signal Processing, vol. 42, no. 1-2, pp. 314–334, 2014.

[2] M. Toothman, K. Barton, B. Braun, S. J. Bury, M. Dessauer, K. Henderson, J. Moyne, and R. Wright, "Trend-based assessment of industrial rotating equipment health," in 2020 Virtual AIChE Annual Meeting. AIChE, 2020.

[3] M. Toothman, B. Braun, S. J. Bury, M. Dessauer, K. Henderson, D. M. T. Ray Wright, J. Moyne, and K. Barton, "Trend-based repair quality assessment for industrial rotating

equipment," IEEE Control Systems Letters, vol. 5, no. 5, pp. 1675–1680, 2020.

[4] Y. Zhao, M. Toothman, J. Moyne, and K. Barton, "An adaptive modeling framework for bearing failure prediction," Electronics, vol. 11, no. 2, p. 257, 2022.

[5] L. R. Rabiner, "A tutorial on hidden markov models and selected applications in speech recognition," Proceedings of the IEEE, vol. 77, no. 2, pp. 257–286, February 1989.

[6] C. J. Lu and W. O. Meeker, "Using degradation measures to estimate a time-to-failure distribution," Technometrics, vol. 35, no. 2, pp. 161–174, 1993.

[7] W. Krzanowski, "Between-groups comparison of principal components," Journal of the American Statistical Association, vol. 74, no. 367, pp. 703–707, 1979.

53c The Pros and Cons of Machine Learning Versus Physical Model ? Case Study on Hydrocracking

Spagnol Adrien, Benoit Celse, Gabriel Ducret, Per Julian Becker

Transformation of heavy crude oil into more valuable components: light (naphtha) and middle distillates (kerosene and diesel) is becoming increasingly important for the refining industry. Hydrocracking plays a major role in this transformation. Development of models for industrial hydrocrackers has received a great amount of attention by the scientific community over the past decades. These models can be used for several targets: chose optimal operating condition in order to optimize one target (Real Time Optimization), carry out what if analysis in order to estimate the impact of feed variations... Depending on the target and the reactor to model (Hydrotreating or Hydrocracking), several kinetic models can be used from very simple (classical lumping) to much more complex ones (single events).

These models are challenged by new techniques such as machine learning or deep learning.

This paper proposes a global methodology including design of experiments, deactivation model, Yields models and product properties models. It indicates the best models to use depending on the target (either first principles or machine learning models). And the pros and cons of Machine Learning. Some examples from hydrotreating or hydrocracking will be provided with comparison between first principles models and machine learning models when reliable.

This global methodology can be used for any kind of process.

131b Applying Multivariate Modeling and Numerical Optimization to Two Product Development Datasets

Marlene Cardin, Alexander Nguyen

Multivariate modeling (PLS) is a proven modeling method that is known to handle the highly correlated variables and missing data challenges often present in product development datasets. Additionally, datasets of various sizes can be effectively modeled and inverted using constrained numerical optimization. In this presentation, both PLS modeling, simulation, and numerical optimization will be applied to two datasets with the goal of developing new or reformulating existing products. These two applications include:

- Development of new LDPE resins with targeted properties
- EPDM Rubber compound adhesion modeling and design of targeted experiments

In both datasets the following will be highlighted with the use of ProSensus' FormuSense software:

1. The appropriate modeling structure given the data available
2. Modeling results and model limitations due to dataset limitations
3. The design of new experiments or products through simulation or optimization

131c Advanced Digital Data Capture and Analysis Tools Replacing Traditional Visual Performance Ratings of Coatings

Yicheng Hu, Kevin Henderson, Jonathan DeRocher, Sun Hye Kim, Allyson Marianelli, Alicia Liew, Nipun Bisht, Maxwell Ingram, Michael Tran, Yang Liu, Ce Xi, Michael Linsen, Mathias Weeks

Many critical coating performance attributes are traditionally rated by researchers by using a visual assessment to quantify and predict film performance. The rating is left up to the researcher to assess in a subjective matter, often using guidelines provided by an industry specification or third parties. New digital image capture technology coupled with controlled lighting equipment eliminates inconsistencies of human visual perception and enables accurate and consistent means to capture and digitize the appearance of coatings. However, digital data capture alone does not address the complexity of rating performance and being able to compare with reference data.

Working toward a holistic digital infrastructure, the Dow Coating Materials team has developed novel processing algorithms used in conjunction with standardized imaging stations to revolutionize the testing of traditionally subjective visual methods for coatings. Through customization of both the hardware and software interface, images are captured in a consistent format, tagging relevant experimental context and metadata. Sophisticated processing algorithms leverage the images to produce robust, quantifiable analysis and ratings devoid of user subjectivity. The resultant data, stored in a centralized database, is available for future access and as a historical record for future algorithm development and improvement. Here we discuss the use of a digital infrastructure for image analysis in select coatings applications, including corrosion resistance and exterior exposure. The deployment of such digital tools improves the consistency and rigor of technical approaches dependent on visual metrics that are ubiquitous in the coatings industry.

143a A Perspective of AI, Machine Learning and Data Science Towards Industry 4.0

Hong Zhao, Ashok Rao, John Campbell, Jason You

The current excitement about artificial intelligence (AI), particularly machine learning (ML), industrial internet of things (IIoT), Industry 4.0 is palpable and contagious. The expectation that AI is poised to "revolutionize," perhaps even take over, humanity has elicited prophetic visions and concerns from some luminaries. Sure, the advances AI has made in the last 10 years, for example, AlphaGo, autonomous cars, Alexa, Watson, and other such systems, in game playing, robotics, computer vision, speech recognition, and natural language processing are indeed stunning advances. But, as with earlier AI breakthroughs, such as expert systems in the 1980s and neural networks in the 1990s, there is also considerable hype and a tendency to overestimate the promise of these advances. So, what are different this time? What are the true potential and values for industry?

We would like to share some views and perspective through a review of recent innovative advances of system identification, model predictive control, and real-time data analytics in process industry applications. More importantly, we explore how those innovations and applications successfully added tremendous value for process industries. As industrial practitioners of AI/ML, we witnessed the progressive development in applications of AI/ML technology for advanced process control (APC), process optimization, and asset performance management (APM), also learnt lessons of what is working and what not. AI/ML is a powerful enabler, but success requires a solid foundation of domain knowledge and understanding of both AI/ML and the use cases to address. On this road, challenges and opportunities exist simultaneously, we propose a list of strategies for industrial applicants references.

143b A practitioner's assessment of Deep Reinforcement Learning for industrial process control applications

Thomas Badgwell, R Donald Bartusiak

Reinforcement Learning (RL) is a machine learning technology in which a computer agent learns, through trial and error, the best way to accomplish a particular task [1]. The recent development of Deep Reinforcement Learning (DRL) technology, in which deep learning networks [2] are used to parameterize the RL agent's policy and value functions, has enabled superhuman performance for some tasks, most especially games such as chess and GO [3]. Researchers have devised DRL algorithms specifically for continuous control problems [4,5], and this has led many in the process control community to wonder what impact DRL technology will have on our industry. Two recent papers provide an introduction to DRL technology and an analysis of its appropriateness for industrial process control applications [6,7]. This presentation builds on the previous analysis efforts by proposing a that an industrial process control technology must be:

- Intelligent
- Consistent
- Offset-free
- Nominally stable
- Flexible

DRL technology will be assessed with respect to each requirement, exposing existing deficiencies, useful research directions, and potential solutions. It will be shown that DRL technology is not likely to replace currently used Proportional-Integral-Derivative (PID) or Model-Predictive Control (MPC) algorithms, however it may prove useful in managing control systems by helping to tune controllers [8,9,10], advising operators during upsets or transient operations [6], and by managing plant-wide disturbances such as diurnal changes and weather events [6].

A final section will outline several promising research directions for DRL technology [11].

References

[1] RS Sutton, AG Barto, "Reinforcement Learning – An Introduction", The MIT Press, (2018).

[2] I Goodfellow, Y Bengio, A Courville, "Deep Learning", The MIT Press, (2016).

[3] D Silver, A Huang, CJ Maddison, A Guez, L Sifre, G Van Den Driessche, J Schrittwieser, I. Antonoglou, V Panneershelvam, M Lanctot, "Mastering the game of Go with deep neural networks and tree search", Nature 529, 484–489 (2017).

[4] TP Lillicrap, JJ Hunt, A Pritzel, N Heess, T Erez, Y Tassa, D Silver, D Wierstra, "Continuous control with deep reinforcement learning", arXiv:1509.02971 (2015).

[5] L Busoniu, T deBruin, D Tolic, J Kober, I Palunko, "Reinforcement learning for control: Performance, stability, and deep approximators", Annual Reviews in Control, 46, 8-28, (2018).

[6] J Shin, TA Badgwell, KH Liu, JH Lee, "Reinforcement Learning – Overview of recent progress and implications for process control", Computers & Chemical Engineering, 127, 282-294 (2019).

[7] RN Nian, J Liu, B Huang, “A review on reinforcement learning: Introduction and applications in industrial process control”, Computers & Chemical Engineering 139, 106886 (2020).

[8] T Badgwell, K Liu, N Subrahmanya, M Kovalski, “Adaptive PID controller tuning via deep reinforcement learning”, US Patent 10,915,073, (2021).

[9] O Dogru, K Velswamy, F Ibrahim, Y Wu, AS Sundaramoorthy, B Huang, S Xu, M Nixon, N Bell, “Reinforcement learning approach to autonomous PID tuning”, Computers & Chemical Engineering, 161, 107760 (2022).

[10] NP Lawrence, MG Forbes, PD Loewen, DG McClement, JU Backstrom, RB Gopaluni, “Deep Reinforcement Learning with Shallow Controllers: An Experimental Application to PID Tuning”, arXiv, 2111.07171v1 (2021).

[11] A Mesbah, KP Wabersich, AP Schoellig, MN Zeilinger, S Lucia, TA Badgwell, JP Paulson, “Fusion of Machine Learning and MPC under Uncertainty: What Advances Are on the Horizon?”, American Control Conference, June 8-10, Atlanta, GA (2022).

143c Framework for Hybrid Machine Learning with Open-Source Python Seeq Sysid Package

Mohammad R. Babaei, Junho Park, Ashwin N. Venkat, John Hedengren

Increased computational resources and machine learning methods have triggered a new era of data science that has transformative potential across many fields, but these possibilities have yet to come to fruition because physics-based and other a priori information is not incorporated into the current machine learning algorithms. This work presents a new open-source Seeq add-on to identify input and output relationships to create digital twin models for various analytics and machine learning tasks. The System Identification Add-on supports the construction of a variety of dynamic models. White box models are primarily based on physics-based principles such as conservation laws and reaction kinetics for the chemical process. Development of these models requires extensive domain knowledge and requires domain knowledge to build and maintain. Machine learning models are largely data-driven. Greybox models combine the strengths of both approaches; they simplify the white-box model by lumping several parameters into fewer parameters that can be identified from data with higher degrees of confidence. These models are capable of capturing the primary dynamics of the system but are valid over larger ranges of operation than purely data-driven models. Currently supported model identification options are ARX, ARIMAX, FIR, transfer function, subspace, and neural network. Identified models can be a blend of time-series, state space (continuous or discrete), nonlinear differential and algebraic equations (DAEs), continuous and integer variables, and machine learned models (Neural network, Gaussian Processes, and Support Vector Regressors) [1] within the Gekko underlying modeling platform [2]. Future model integration development includes constrained SysID [3-5], Transformer Neural Networks [6], Transfer learning, and Physics-Informed Neural Networks [6,7].

The SysID Python package integrates with Seeq Workbench for data exploration and analysis. Cleaned and contextualized data from Seeq workbench is used to identify dynamic models and the models are returned back into Seeq workbench or exported to other environments. This framework supports both batch historical data investigations as well as streaming predictions. The system identification framework is used for a variety of tasks including simulation, digital twin construction, advanced process control, and real-time optimization modeling.

References

[1] Gunnell, L., Manwaring, K., Lu, X., Reynolds, J., Vienna, J., Hedengren, J.D., Machine Learning with Gradient-based Optimization of Nuclear Waste Vitrification with Uncertainties and Constraints, Processes, MDPI, 2022, accepted for publication. https://gekko.readthedocs.io/en/latest/ml.html

[2] Beal, L.D.R., Hill, D., Martin, R.A., and Hedengren, J.D., GEKKO Optimization Suite, Processes, Volume 6, Number 8, 2018, doi: 10.3390/pr6080106.

[3] Park, J., Lyman, J., Darby, M., Lima, L., Nelson, C., and Hedengren, J.D., Hybrid Machine Learning and Fundamental Modeling for Real-Time Optimization of a Fluidized Bed Roaster, 2020 Spring Meeting & 16th Global Congress on Process Safety, AIChE, Houston, TX, 29 March-2 April, 2020.

[4] Park, J., Hansen, B., Gates, N., Darby, M., Hedengren, J.D., Use of Nonlinear and Machine Learning Techniques for Improved APC Modeling, AIChE Spring Meeting, New Orleans, LA, April

2019.

[5] Nikbakhsh, S., Hedengren, J.D., Darby, M., Udy, J., Constrained Model Identification Using Open-Equation Nonlinear Optimization, AIChE Spring Meeting, Houston, TX, April 2016.

[6] Park, J., Hybrid Machine Learning and Physics-based Modeling Approaches for Process Control and Optimization, Ph.D. Dissertation, Brigham Young University, 2022.

[7] Knotts, T., Hedengren, J.D., Babaei, M.R., Physics-Informed Deep Learning for Prediction of Thermophysical Properties: The Parachor Method for Surface Tension, AIChE Annual Meeting, Phoenix, AZ, Nov 13-18, 2022.

AUTHOR INDEX

AIChE
120 Wall Street, FL 23
New York, NY 10005-4020

CURRAN ASSOCIATES INC.
proceedings.com

ISBN 978-1-7138-7935-0